Laboratory Science with Space Data

Foreword by Giuseppe Reibaldi

Space exploration was initiated only 50 years ago but it has already given mankind a different perspective on the way we look at the universe and at our position within it.

Human spaceflight is the pinnacle of space exploration, and the International Space Station (ISS), with its global international participation and its outstanding achievements, tangibly demonstrates what can be accomplished. The ISS Programme is the most complex space project ever undertaken, with 15 countries involved. Its construction began in 1998 and was completed in 2010, with a total mass of about 400 metric tons and the size of a football field.

The European Space Agency (ESA) joined the ISS programme in 1995, with several projects; among those, the most important contribution is the Columbus Laboratory, a multi-user research outpost orbiting the Earth at an altitude of 400 km, with a speed of 28,000 km/h, in a permanent state of low gravity.

The Columbus Laboratory, attached to the ISS since 2008, is outfitted with a variety of multi-user scientific equipment to conduct investigations in different fields, from human physiology to biology, materials science, fundamental physics and others. Columbus is planned to be in operation until at least 2020.

Gravity is the only physical parameter that has never changed since the Earth was formed and investigations conducted in low gravity conditions, such as those offered by the Columbus Laboratory, are expected to lead to major discoveries, which will make a key contribution to the innovation cycle crucially needed in Europe.

Indeed, the availability of open, competitive and quality-based access to pan-European global research infrastructures is becoming increasingly important at European Union (EU) level. Resolutions adopted at meetings of the Space Council have advocated a strong synergy between the EU Framework Programmes and ESA activities; the European Parliament has also formulated a similar encouragement.

In line with the above, ESA has initiated actions aiming to integrate the Columbus Laboratory, as a unique multi-user European research infrastructure in orbit, into the European Research Area. The European segment of the ISS including

Columbus can be defined as the European Laboratory in Space, and the research conducted onboard will have a multiplier effect on the R&D budgets of the EU.

An early example of the advocated synergy between ESA and the EU in the field of human space exploration is the ULISSE project, funded by the EU.

The ULISSE project is aiming, among other goals, to set up a system that makes the scientific data generated onboard the ISS easily available to a vast community of researchers, while ESA will generate the scientific data onboard Columbus.

The combination of these activities is considered a fundamental concrete step towards increasing the return on investment for the European population, and playing an important role in responding to the Grand Challenges Europe is facing, as well as in supporting the Europe 2020 Vision strategy.

Estec, Netherlands
January 2011

Giuseppe Reibaldi
Head of the Special Projects Office
Directorate of Human Spaceflight
European Space Agency

Foreword by Carlo Albanese

This book originates from the results of the ULISSE project, which is co-funded by the European Commission within the Seventh Framework Programme.

It is worth noting the context in which the project has arisen and the needs and perspective it is attempting to address: the context of the global knowledge society. Everything is becoming global, this has both benefits and drawbacks; however one area in which this general trend leads only to advantages is the generation of global knowledge and the possibility to share the benefits derived from cooperative scientific research.

Scientific research in Space can be seen as a paradigm of how a common objective can be achieved through a worldwide cooperation, and the building of the International Space Station is one of the clearest examples of that.

Even if access to Space and its use and control remain some of the most widely discussed arguments, due to their direct impact on the security of nations, nevertheless in the field of scientific research in Space it is time that the results obtained up to now became a common heritage for all humanity, to enable citizens to take advantage in their daily life. In this, Europe can be an example to all other nations.

The International Space Station is an exceptional Laboratory, the availability of advanced facilities in a stable space environment with humans on board makes it one of the most prominent research infrastructures available for the European science communities. However, there remains a fundamental gap that still needs to be filled.

This is mainly due to the fact that the results obtained are accessible only to a restricted community of users who became more acquainted with experimentation in microgravity in the last two decades. These results need to be made available to a wider community of researchers, even those not directly involved in space experimentation. They will be able to look at these results from a different perspective enhancing the interdisciplinary approach and supporting research in fields different from the ones in which the experiments were originally conceived. This approach will stimulate new ideas and new possibilities to discover applications with an impact on the daily life of citizens. Not only should space results be exploited to their maximum potential, they also need to be part of the education of new generations in order to establish a common knowledge on which we can build our future.

Today we have in our hands technologies that offer us unprecedented possibilities to integrate and share distributed knowledge, and to develop tools for the analysis, promotion and dissemination of the body of knowledge generated by space experimentation. The ULISSE project intends to apply these technologies so that an ever-growing number of users is able to access data for developing new applications and new research perspectives, on the ground as well as in Space, looking at planetary exploration, and, last but not least, to contribute to promoting a cultural change in the way citizens consider Space – as an environment in which it is possible to work and achieve results that improve our lives.

To accomplish these results, many issues and challenges have been faced, some have been solved, others need further research, but all the people involved have a clear understanding of the relevance of the perspective they are working on, and all are confident that ULISSE will ultimately reach Ithaca.

Naples, Italy
January 2011

Carlo Albanese
Telespazio s.p.a
Head of Science Systems and Applications
Via E. Gianturco 31
80146 Naples – Italy

Contents

1 **Introduction** 1
References 4

2 **Space and Ground-Based Infrastructures** 7
2.1 Characteristics of the Space Environment 7
2.1.1 External Environmental Conditions 8
2.1.2 Internal Spacecraft Environmental Conditions 8
2.2 Space Infrastructure 9
2.2.1 The International Space Station 9
2.2.2 Foton/Bion Recoverable Capsules 30
2.3 Non-orbital Weightless Research Platforms 31
2.3.1 Sounding Rockets 32
2.3.2 Parabolic Flight Airplanes 33
2.3.3 Drop Towers 35
2.3.4 High-Altitude Balloons 36
2.4 Ground-Based Facilities and Space Simulators 37
2.4.1 Bed Rest, Hypokinesia and Metabolic Balance Facilities 37
2.4.2 Isolation and Cônfinement, Pressure Chambers and Climate Chambers 38
2.4.3 Centrifuges 39
2.4.4 Human-Rated Linear and Angular Accelerators 41
2.4.5 Clinostats, Free Fall Machines, and Random Positioning Devices 42
2.4.6 Telemedicine, Behaviour and Metrics 43
2.4.7 Integrated Bio-Processing, Tissue Engineering 44
2.4.8 Magnetic Resonance Facilities 45
2.4.9 Movement Analysis, Physical and Skills Training 45
2.4.10 Additional Animal Physiology Facilities 45
2.4.11 Additional Plant Physiology Facilities 46
2.4.12 Magnetic Levitation 46

2.4.13 Biotechnology and Life Support Systems 47
2.4.14 Extreme Environments 47
2.4.15 Radiation Testing 48
2.4.16 Fluid Science Facilities: Surface Tension 49
2.4.17 Materials Science Facilities: Crystallisation 49
2.4.18 Materials Science Facilities: Solar Power 49
2.4.19 Materials Science Facilities: Wind Tunnels 50
References 50

3 Areas of Research 55
3.1 Physical Sciences 56
3.1.1 Fundamental Physics 56
3.1.2 Fluid Physics by Daniel Beysens 64
3.1.3 Combustion by Christian Chauveau 75
3.1.4 Materials Science 81
3.1.5 Space Weather by Hanna Rothkaehl 97
3.2 Life Sciences 103
3.2.1 Astrobiology 103
3.2.2 Gravitational Cell Biology 107
3.2.3 Radiation Biology 113
3.2.4 Plant Sciences 117
3.2.5 Animal Physiology 123
3.2.6 Human Physiology 129
3.2.7 Psychology 141
3.2.8 Research for Operational Space Medicine 145
3.2.9 Research on Life Support Systems 153
References 157

4 ULISSE Access to Data 171
4.1 Data Description 171
4.1.1 Preservation of Space Data 171
4.1.2 Main Space Databases 172
4.1.3 Preservation and Valorisation Approach 173
4.1.4 Initial Production of Data 174
4.1.5 Data Support and Storage 177
4.1.6 Data Resulting from the Experiments and Processing Levels 178
4.1.7 Data Formats 181
4.1.8 Samples 182
4.1.9 Legal Aspects and Ethical Issues 183
4.2 Identification of Information: The ULISSE Knowledge Base 185
4.2.1 Purpose and Role of Semantic Technologies 185
4.2.2 Topic Maps as a Semantic Data Model 186
4.2.3 The ULISSE Knowledge Base 192

4.3 ULISSE: Data Access Services 196
4.3.1 Services for Obtaining Data 196
4.3.2 ULISSE Services 199
4.3.3 Example of a Searching and Browsing Session 202
References 205

Index 207

Contributors

Jessica Aceto LBMGG University, 4000 Liège, Belgium

Carlo Albanese Telespazio s.p.a., Via E. Gianturco 31, 80146 Naples, Italy

Joan Albiol Universitat Autònoma de Barcelona, Barcelona, Spain

Audrey Berthier Institute for Space Physiology and Medicine, BP 74404, 31405 Toulouse Cedex 4, France

Daniel Beysens CEA-Grenoble and ESPCI-ParisTech, 10 rue Vauquelin, 75005 Paris, France

Johannes Boonstra Cell Biology, Department Biology, University Utrecht, Padualaan 8, 3584 CH, Utrecht, The Netherlands

Silvia Bradamante CNR-ISTM, Institute of Molecular Science and Technology, Via Golgi 19, 20133 Milan, Italy

Luigi Carotenuto Telespazio s.p.a., Via E. Gianturco 31, 80146 Naples, Italy

Christian Chauveau CNRS – INSIS – ICARE, 1C, avenue de la Recherche Scientifique, 45071 Orléans Cedex 02, France

Alexander Chouker University of Munich, Munich, Germany

Gilles Clément International Space University, Strasbourg, France

Alain C. Colige Laboratory of Connective Tissues Biology, University of Liège, Liège, Belgium

Bernard Comet MEDES – Institute for Space Physiology and Medicine, BP 74404, 31405 Toulouse Cedex 4, France

Christine Cornier Centre spatial de Toulouse, 18 avenue Edouard Belin, 31401 Toulouse Cedex 9, France

Volker Damann ESA/European Astronaut centre, Linder Hoehe, 51147 Cologne, Germany

David Damen Space Applications Services, Leuvensesteenweg, 325, B-1932 Zaventem, Belgium

Marco Durante GSI, Biophysics, Darmstadt, Germany

Pascale Ehrenfreund Institute of Chemistry, Leiden University, Leiden, The Netherlands and Space Policy Institute, George Washington University, Washington, DC, USA

Hans Fecht University of Ulm, Institute of Micro and Nanomaterials, Albert-Einstein-Allee 47, 89081 Ulm, Germany

Aloïs Grimbach Microgravity User Support Center MUSC, Linder Höhe, D-51147 Cologne, Germany

Rüdiger Hampp University of Tübingen, Microbiology Institute, Auf der Morgenstelle 1, D-72076 Tübingen, Germany

Ruth Hemmersbach Biomedical Science Support Center, Institute of Aerospace Medicine, German Aerospace Center (DLR), 51147 Cologne, Germany

Bob Hockey Department of Psychology, University of Sheffield, Norton, UK

Eberhard Horn Karlsruhe Institute of Technology (KIT), 76128 Karlsruhe, Germany

Charles A. Lambert Laboratory of Connective Tissues Biology, University of Liège, Liège, Belgium

Victoria Lapuerta Universidad Politécnica de Madrid, E.T.S.I. Aeronáuticos Plaza de Cardenal Cisneros, 3 28040 Madrid, Spain

Natalie Leys Unit for Microbiology, Institute for Environment, Health and Safety, Belgian Nuclear Research Center SCK/CEN, Boeretang 200, 2400 Mol, Belgium

Cesare Lobascio Thales Alenia Space, Rome, Italy

Jack J.W.A. van Loon DESC @ OCB-ACTA, University of Amsterdam and VU-University, Gustav Mahlerlaan 3004, 1081 LA Amsterdam, The Netherlands

Stefano Mancuso LINV – Department of Plant, Soil & Environmental Science, University of Florence, Viale delle idee 30, 50019 Sesto Fiorentino, FI, Italy

Francisco Javier Medina Centro de Investigaciones Biológicas (CSIC), Ramiro de Maeztu 9, 28040 Madrid, Spain

Giovanni Minei Telespazio s.p.a, Via E. Gianturco 31, 80146 Naples, Italy

Greg Morfill Max-Planck-Institut für extraterrestrische Physik, Giessenbachstrasse, 85748 Garching, Germany

Marc Muller LBMGG, University, 4000 Liège, Belgium

Betty V. Nusgens Laboratory of Connective Tissues Biology, University of Liège, Liège, Belgium

Diana Pasquariello Telespazio s.p.a., Via E. Gianturco 31, 80146 Naples, Italy

Maria Antonietta Perino Thales Alenia Space, Rome, Italy

Günther Reitz DLR, Radiation Biology, Cologne, Germany

Hanna Rothkaehl Space Research Center Polish Academy of Sciences, Bartycka 18A, 00-716 Warsaw, Poland

Gro Mjeldheim Sandal Department of Psychosocial Science, University of Bergen, Bergen, Norway

Laurence Vico University of Saint-Etienne, Saint-Etienne, France

Jon Weems ESA/ESTEC, Noordwijk, Netherlands

Frances Westall Centre de Biophysique Moléculaire-CNRS-OSUC, Rue Charles Sadron, 45071 Orléans cedex 2, France

Martin Zell ESA/ESTEC, Noordwijk, Netherlands

Chapter 1
Introduction

Luigi Carotenuto

Abstract This chapter introduces the context, objectives and structure of the book. This book aims both to contribute to disseminate the knowledge about the scientific research conducted in space and to promote new exploitation of existing data in this field. While space experiments are characterised by a long time for preparation, high costs and few opportunities, significant scientific value is expected from the resulting data for almost scientific disciplines. In this context, ISS is a unique experimental environment for research. As part of its Seventh Framework Programme, the European Commission intends to support further exploitation and valorisation of space experimental data. This book was realised as part of the ULISSE project, co-funded by the European Union. The book intends to provide an introduction to space research with a focus on the experiments performed on the ISS and related disciplines. The book also intends to be a useful guide, not only for scientists but also for teachers, students and newcomers to space research activities. Chap. 2 is dedicated to space environment and infrastructures, Chap. 3 to the main disciplines involved in space research and Chap. 4 on how to get access to space experiment data through ULISSE.

This book has the ambitious objective of contributing to the dissemination of knowledge about the scientific research that is conducted in space, that exploits the unique conditions of that environment such as the vacuum, radiation, confinement and particularly microgravity.

Exploitation of the conditions of the space environment for scientific research began with the Sputnik-2 mission in 1957 and continued in the successive decades [1, 2, 3, 4 as examples]. Since then, it has been clear that experiments conducted in space can lead to unexpected and important breakthroughs in many scientific disciplines, due to the opportunity of observing new phenomena that are masked by gravity on the ground, or of obtaining data of unprecedented accuracy by avoiding the disturbances induced by gravity.

Indeed, space research has already led to a number of important advances in different scientific fields, ranging from properties and behaviour of condensed

D. Beysens et al. (eds.), *Laboratory Science with Space Data*,
DOI 10.1007/978-3-642-21144-7_1, © Springer-Verlag Berlin Heidelberg 2011

matter to a better understanding of bone demineralization and muscle atrophy processes.

On the other hand, experiments performed in space face a number of difficulties:

- preparation takes a long time (typically several years) during which adequate equipment qualified to operate in the difficult conditions of the space environment is designed and developed;
- costs are generally high, due to the huge investments required for space missions;
- opportunities for repeating and improving successful experiments are limited.

For the above reasons, space experiments are selected by the space agencies according to rigorous peer review. It is then expected that the data, information and knowledge produced by these experiments will acquire a very relevant scientific value.

Moreover, since early 2000 the International Space Station (ISS) has been operative [5]. The ISS is currently the largest international space programme: it represents the unique opportunity of a permanent orbital laboratory with six crew members performing technological and scientific experiments on-board. Due to the large increase in operative time and the performance improvement of the instrument technologies compared to those used in earlier space missions, the ISS is leading a new phase of space experimentation, characterised by a quantitative and qualitative increase in its scientific relevance.

In order to support the scientific operations on board the ISS, the European Space Agency (ESA) has established a distributed infrastructure, comprising the European Mission Control Centre and a network of User Support and Operations Centres (USOCs). USOCs have been established in various EU countries with the support of national space agencies and ESA to conduct the operations of European scientific experiments on board Columbus and other ISS modules. For this purpose the USOCs use an ad-hoc infrastructure, dedicated to space operations only. Through the space infrastructure the USOCs receive data and exchange commands with the scientific payloads on board the ISS. During operations the USOCs send data to the scientific investigator(s) responsible for the experiments; when the experiments have been performed data are also archived in the USOC operations infrastructure. However, long-term archiving and preservation of the detailed knowledge needed for data interpretation for the benefit of future user communities is difficult. It is worth noting that the interpretation of space data usually also needs knowledge of environmental conditions and housekeeping data from space equipment.

Data analysis is performed by the investigators responsible for the experiments, who disseminate the results through conferences and publication of papers in scientific journals. However, these usual means of dissemination are targeted at a specialised audience and do not adequately support cross-fertilization among different scientific domains.

Moreover, scientific data produced by experiments on the ISS cannot be accessed straightforwardly, as the ISS operations infrastructure is protected for security reasons. Any scientist interested in re-analysing existing data has to contact

ESA or the investigator, once these reference persons have been identified, on a case-by-case basis.

On the other hand, the scientific value of space data, the time and money spent on their acquisition, and the limited repeatability of space experiments demand that special attention be paid to the results obtained in space.

Following the strategy endorsed by the Lisbon European Council in 2000, the European Union is pursuing the development of a knowledge-based, innovative and dynamic society in Europe; the key to attaining this ambitious goal is the knowledge triangle: research – innovation – education. The Research Framework Programme is the European Union's main tool for promoting research in almost all scientific disciplines, in particular the Seventh Framework Programme (FP7) which covers the period from 2007 to 2013.

The European Union has included the space theme in this programme. This acknowledges the strategic role of space research due to its contribution, especially through its direct applications, to the construction of Europe and the competitiveness of the European Union [6]. As part of the specific FP7 work programme on the space theme [7], the European Commission intends to support the exploitation and valorisation of space experimental data. The ULISSE (USOCs knowLedge Integration and dissemination for Space Science and Exploration) project, involving the entire USOC network, research institutions and organizations like ELGRA (European Low Gravity Research Association), was proposed in the context of the first call of 2007 and the project was accepted for co-funding by the EU (FP7 Grant Agreement no. 218815).

ULISSE is based on the particular expertise and competencies of the USOC network. Each USOC collects detailed knowledge of the scientific payloads for which it is responsible; related operations are conducted in close cooperation with investigators, a deeper understanding is then gained of the relevant scientific aspects of the experiments; finally, each USOC becomes custodian of the data and information about the experiments operated under the USOC's responsibility.

The purpose of ULISSE is to valorize this huge amount of data and information and to support its further exploitation. ULISSE plays a pathfinder role for many issues related to exploitation and dissemination of ISS data. A fruitful collaboration has been established with ESA, which is supporting the implementation of adequate processes for exploiting the data produced by the experiments performed on the ISS, while ensuring protection of Intellectual Property Rights or crew personal (medical) integrity.

The project has characterized a reference scenario for space data exploitation, ranging from the definition of user needs and required services, to the identification of existing data sources and related data policies and dissemination constraints. In ULISSE the additional information needed for future data exploitation has also been identified and described in a specific metadata standard.

In addition to these exploratory studies, ULISSE has been pursuing the realization of a demonstrator, which uses a representative subset of the identified functionalities; tools and services are accessible through an internet portal. A network has been implemented to ensure connectivity between the distributed resources; a number of tools have been developed to interface the distributed repositories located at the premises of the different USOCs and other data providers; additional tools and applications have been developed to support data exploitation. Special attention has been devoted to the generation and management of metadata and to knowledge representation and management. For knowledge representation, innovative technologies have been used to promote associations and consequently potential cross-fertilization among different scientific domains. The ULISSE tools provide access to the knowledge base and to metadata from all kinds of experiments, with links to the required data according to the applicable data policy. Additional tools have been developed for data integration, graphical three-dimensional data display and to support the planning and validation of activity sequences for scientific space experiments. ULISSE also includes a promotional portal to improve the dissemination of results of space research and cooperation among the interested user communities.

This book, in keeping with European Union guidelines about dissemination of research results, intends to provide an introduction to space research with a focus on the experiments performed on the ISS and related disciplines; aimed at members of the physics, life sciences, medicine and engineering communities. The book also intends to be a useful guide, not only for scientists but also for teachers, students and newcomers to space activities, to approaching research in space, the scientific disciplines involved and the data and knowledge infrastructure provided by ULISSE.

Chapter 2 describes the main features of the space environment and their relevance for research. It also discusses the orbital and sub-orbital platforms that are used for space research. Finally a survey is provided of the ground-based facilities and space simulators that provide complementary data.

Chapter 3 presents a survey of the main disciplines involved in space research, ranging from physical science and engineering (which include fundamental physics, fluid and material sciences, combustion and space weather) to life sciences (which include human and animal physiology, gravitational and radiation biology, astrobiology, plant sciences, space medicine and life support systems).

Chapter 4 illustrates some of the contributions of the ULISSE project to the exploitation of scientific space data. The life cycle of space data is described, together with the main constraints and legal aspects influencing data access and dissemination. The approach and the technology adopted for establishing the ULISSE knowledge base are outlined; and finally the main ULISSE services are described, with an example of utilization as a guideline for users.

References

1. Antipov, V.V., Kozlov, V.A., Parfenov, G.P., Saksonov, P.P.: The results of biological studies made on board the Voskhod and Voskhod 2 spaceships. Life Sci. Space Res. **5**, 119–121 (1967)
2. Apollo-Soyuz Test project: Summary Science Report, NASA SP-412 (1977)

3. Gazenko, O.G., Giurdzhian, A.A.: Medical and biological studies on the cosmic ships of the "Voskhod" type. Vestn. Akad. Nauk SSSR **35**(8), 19–26 (1965)
4. Edwards, B.F.: Weightlessness experiment on Biosatellite II. Life Sci. Space Res. **7**, 84–92 (1969)
5. SPACE Base Europe: ESA BR 270, ISBN 978-92-9221-006-9 (2007)
6. Communication from the Commission to the Council and the European Parliament: European Space Policy: COM (2007) 212 final, 26 April 2007
7. European Commission Decision C (2010) 4900 of 19 July 2010

Chapter 2
Space and Ground-Based Infrastructures

Jon Weems and Martin Zell

Abstract This chapter deals first with the main characteristics of the space environment, outside and inside a spacecraft. Then the space and space-related (ground-based) infrastructures are described. The most important infrastructure is the International Space Station, which holds many European facilities (for instance the European Columbus Laboratory). Some of them, such as the Columbus External Payload Facility, are located outside the ISS to benefit from external space conditions. There is only one other example of orbital platforms, the Russian Foton/Bion Recoverable Orbital Capsule. In contrast, non-orbital weightless research platforms, although limited in experimental time, are more numerous: sounding rockets, parabolic flight aircraft, drop towers and high-altitude balloons. In addition to these facilities, there are a number of ground-based facilities and space simulators, for both life sciences (for instance: bed rest, clinostats) and physical sciences (for instance: magnetic compensation of gravity). Hypergravity can also be provided by human and non-human centrifuges.

2.1 Characteristics of the Space Environment

Space is a very demanding set of characteristics [18] when we think of designing experiments and associated equipment to function in this environment. Conditions vary greatly, not only within the universe and our solar system but also around our planet. Factors such as temperature and radiation levels vary greatly at different altitudes and under the influence of Earth's magnetic field. We will consider the conditions experienced by spacecraft in low-Earth orbit as this is a principal orbital location where space research is undertaken. For explanatory purposes we will consider the International Space Station (ISS) as the orbital spacecraft.

D. Beysens et al. (eds.), *Laboratory Science with Space Data*,
DOI 10.1007/978-3-642-21144-7_2, © Springer-Verlag Berlin Heidelberg 2011

2.1.1 External Environmental Conditions

The ISS orbits the Earth approximately every 90 min (16 times per day) at around 350–400 km above the surface of the Earth. Outside the Station there is a high quality vacuum (minimum of 3.6 × 10–11 kPa) with a temperature variation from −120 to +120°C. Earth's low-density residual atmosphere at this altitude is primarily composed of atomic oxygen which can cause erosion of certain surfaces. Ionising radiation results from trapped electrons and protons as well as solar and galactic cosmic rays, but also heavier atomic nuclei. These particles can cause degradation and changed states in electronic devices and materials. Of particular concern are solar flares, which temporarily create an intense environment of protons and heavy ions. The trapped proton environment around the Earth dips to only a couple of 100 km above the South Atlantic, creating the 'South Atlantic Anomaly'. On half of its daily orbits the ISS will pass through this trapped proton belt for a duration of typically 5–10 min. The proton flux is then several orders of magnitude higher. With respect to electromagnetic radiation the highest power densities irradiating the ISS are from solar radiation in the ultraviolet and visible portions of the electromagnetic spectrum. Plasma is another factor present in low-Earth orbit. This quasi-neutral gas consisting of neutral and charged particles causes electrical charge accumulation when interacting with a spacecraft until electrical equilibrium is reached. Active components such as solar arrays may accumulate sufficient negative potential to produce arcing to other spacecraft elements [64].

2.1.2 Internal Spacecraft Environmental Conditions

The quality of weightlessness in spacecraft is an important factor for low-Earth orbit research and the ISS residual gravity level is to all intents and purposes a weightless environment. This is determined/affected by atmospheric drag from residual atmosphere, as well as higher magnitude vibrations caused by ISS system activity, crew movement, spacecraft docking/undocking and thruster firings etc. though ISS activities are planned to avoid unnecessary vibrational influence on experiments. This influence is further attenuated by the use of different vibration isolation systems within racks and facilities.

The pressure inside the ISS is kept at around 1 bar or ~100 kPa (*i.e.* normal barometric pressure at sea level) and the air composition is 78% nitrogen and 21% oxygen with a residual carbon dioxide level, which is higher than on Earth (0.05%), though kept within safe limits for crew safety by the use of CO_2 scrubbers. 1% CO_2 is allowable during crew exchanges though the ISS programme has agreed to maintain the cabin CO_2 level to 0.37% (with the goal of reaching 0.3%) for two 90-day periods each year. The temperature on board varies from 17 to 28°C, and a relative humidity varies between 25% and 75%. A residual radiation environment roughly 50–100 times higher than on Earth at sea level does exist inside the ISS, though this is within levels considered safe for the duration of stay of crew members on the ISS.

2.2 Space Infrastructure

Europe has been a strong proponent of the utilisation of space for research purposes for many decades, however, in order to undertake major programmes in space utilisation it became a necessity to form a European-wide agency in the 1960s. This started as the European Space Research Organisation (ESRO), which was founded by ten European nations. In 1975 ESRO together with the European Launch Development Organisation (ELDO) merged to form the European Space Agency (ESA) [58] which now has 18 Member States from across Europe. Prior to this, one of the major programmes agreed upon by ESRO in 1973 together with NASA was Spacelab, the European-built pressurised laboratory module that would fly on 25 different Shuttle flights from 1983 to 1998. This major milestone in European research infrastructure development has been a stepping stone to the development of the major European pressurised modules that are core elements of the International Space Station (ISS) and European utilisation activities in space.

2.2.1 The International Space Station

The ISS (Fig. 2.1) is one of the most extensive civil engineering projects ever undertaken. It is the principal platform for scientists and technology developers worldwide to gain access to the environmental conditions in space for undertaking

Fig. 2.1 The International Space Station photographed from STS-132 space shuttle Atlantis after shuttle undocking in May 2010 (Source: NASA)

research projects across all scientific domains and providing an important tool for education/public relations. The Station [69] is a co-operative programme between United States, Russia, Canada, Japan and 11 Member States of ESA (Belgium, Denmark, France, Germany, Italy, Netherlands, Norway, Spain, Sweden, Switzerland and the United Kingdom). It is governed by an international treaty [55] signed by these International Partners which provides the framework for design, development, operation, and utilisation of a permanently inhabited civil Space Station for peaceful purposes. Utilisation rights are outlined in Memoranda of Understanding. ESA's allocation rights comprise 8.3% of the Space Station's non-Russian utilisation resources including, 8.3% of crew time. In compensation for providing resources (energy, robotics, cooling, telecommunications, etc.) to the European ISS Columbus Laboratory by the National Aeronautics and Space Administration (NASA) and the Canadian Space Agency (CSA), Europe provides 49% of the laboratory's utilisation resources to NASA and 2% to the CSA (via NASA).

2.2.1.1 The European Columbus Laboratory

Europe carried out almost 200 experiments on the ISS from 2001 until the launch of Columbus in February 2008 mainly associated with short-duration missions involving European astronauts. The related science packages reflected specific nationally sponsored Soyuz taxi flight missions from France, Italy, Belgium, Spain, and the Netherlands, with German ESA astronaut Thomas Reiter undertaking Europe's first long-duration (6-month) ISS mission in 2006. Columbus is ESA and Europe's biggest single contribution to the ISS and its arrival greatly increased Europe's research potential. Along with the European-built ISS Nodes 2 and 3, it shares its basic structure and some system elements with the Italian Space Agency's (ASI) Multi-Purpose Logistics Modules, one of which now constitutes a permanent ISS storage module called the Permanent Multi-Purpose Module (PMM).

The Columbus laboratory (Fig. 2.2) is equipped with a suite of flexible multi-user facilities that offer extensive research capabilities. During its lifespan Earth-based researchers, with the assistance of the ISS crew and an integrated network of ground control centres will be able to conduct many experiments in physiology, biology, materials science, fluid physics and a host of other disciplines. Columbus can host ten standard-sized research racks [3] which are provided with power, cooling, video and data lines. ESA developed a range of research racks, principally located in Columbus to offer European scientists full access to a weightless environment that cannot be duplicated on Earth.

During compilation of this publication the configuration of Columbus included the following ESA research rack facilities.

- **The Fluid Science Laboratory (FSL)** – accommodating experiments looking into fluid flow, heat transfer and foam stability/instability mechanisms in weightlessness, which could bring far-reaching benefits on Earth such as more efficient oil extraction processes.

Fig. 2.2 European Columbus Laboratory photographed in June 2008 during the STS-124 mission (Source: NASA)

- **The European Drawer Rack (EDR)** – a modular and flexible facility providing basic accommodation and resources for standard sized experiment modules covering a large variety of scientific disciplines,.
- **Biolab** – supporting experiments on micro-organisms, cells and tissue cultures, small plants and small animals (insects, worms).
- **The European Physiology Modules Facility (EPM)** – used to investigate the effects of long-duration spaceflight on the human body and contributing to an increased understanding of conditions such as osteoporosis and balance disorders on Earth as well as providing an insight into neurological mechanisms.
- **Muscle Atrophy Research and Exercise System (MARES)** – used for undertaking neuromuscular and exercise research on the International Space Station by assessing the strength of isolated human muscle groups around joints.

In addition to these European research racks Columbus hosts: two NASA Human Research Facility (HRF) racks for supporting physiology research; an additional NASA EXPRESS rack which incorporates ESA's European Modular Cultivation System (EMCS) for undertaking additional biological research; and the European Transport Carrier which acts as a stowage rack for supporting European research. The final rack location in temporarily Columbus housed the European-built Microgravity Science Glovebox (MSG) for undertaking a variety of materials, combustion, fluids and biotechnology experiments, though this was relocated back to the US Laboratory.

In addition to stand-alone elements of European research hardware in orbit, the full spectrum of major European research facilities inside the ISS is completed by the **Material Science Laboratory (MSL)**, which is the principal element of NASA's **Material Science Research Rack (MSRR-1)**, also in the US laboratory.

The external surface of the Space Station offers great potential for undertaking a full spectrum of exposure research and technology demonstrations in for example astrobiology, materials science, astrophysics and Earth observation. Columbus offers four external payload locations exposed to space vacuum [4], with an unhindered view of Earth and outer space, and supplied with relevant resources (power, data). One Columbus external payload, the **European Technology Exposure Facility (EuTEF)** already returned to Earth following completion of a successful 1½ years exposure to open space in September 2009, and the **SOLAR** facility, which was also installed outside of Columbus in February 2008 remains on orbit until 2013 and possibly beyond.

With the extensive amount of research facilities on the ISS, undertaking research on the Station only involves the transport of new external payloads, experiment samples and/or ancillary equipment in many cases. This is brought to the ISS by either one of the unmanned logistics vehicles (European ATV, Japanese HTV or Russian Progress) or by one of the manned ISS vehicles (Russian Soyuz-TMA or in the past namely by the US Space Shuttle). In coming years other platforms such as the SpaceX capsule will also be used for downloading samples to Earth.

2.2.1.2 Europe's Research Facilities in Columbus

The Biolab facility [31] is designed to support biological experiments on micro-organisms, cells, tissue cultures, small plants and small invertebrates. The major objective of performing life science experiments in space is to identify the role that weightlessness plays at all levels of an organism, from the effects on single cells up to complex organisms including humans.

The biological samples, together with their ancillary items are transported from the ground to Biolab either already integrated within dedicated Experiment Containers or in small vials. The latter case applies if the samples require storage prior to use. On-orbit, the Experiment Containers are manually inserted into Biolab (Fig. 2.3 left) for processing, whereas the frozen sample will first be thawed out inside the BioGlovebox. Once this manual loading is accomplished, the automatic processing of the experiment can be initiated by the crewmember.

The experiments are undertaken in parallel on a microgravity and a 1 g centrifuge. The latter provides the flight reference experiment, whilst the ground reference experiment is performed at the Facility Responsible Centre (Microgravity User Support Centre in Cologne, Germany), which operates the facility according to the needs of individual Experiment Container providers. During processing of the experiment, the facility handling mechanism transports the samples to the facility's diagnostic instrumentation, where, through teleoperations, the scientist on the ground can monitor the processing of their experiments from their own User Home Bases and actively participating in preliminary in-situ analyses. Typical experiment durations range from 1 day to 3 months. One example of an experiment undertaken in Biolab is the Waving and Coiling of Arabidopsis Roots (WAICO) experiment.

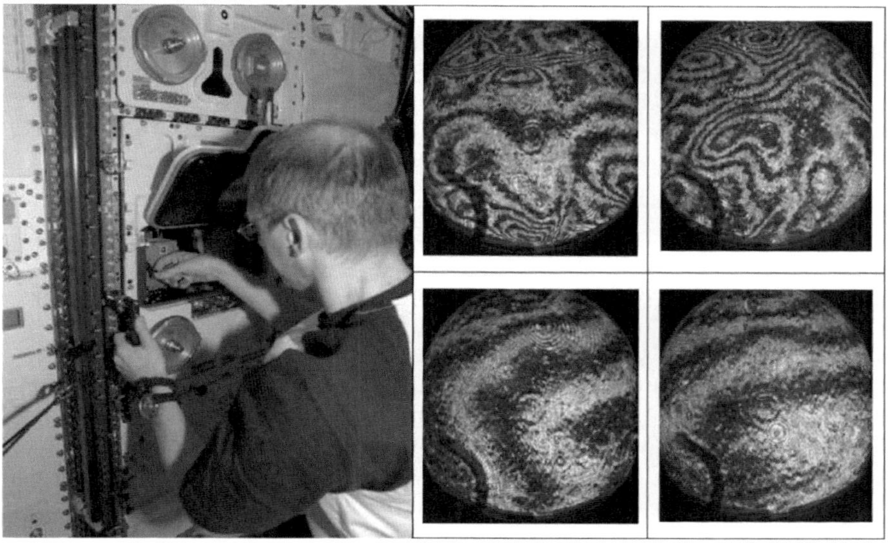

Fig. 2.3 *Left*: ESA astronaut Frank De Winne installing the Yeast experiment in the Biolab in October 2009 (Image: NASA). *Right*: Geoflow interferometric image from ESA's Fluid Science Laboratory on board the Columbus laboratory (Source: Prof. C. Egbers, BTU Cottbus)

Fluid Science Laboratory – European Rack Facility for Fluids Research on the ISS

The Fluid Science Laboratory [32] is used for studying the dynamics of fluids in the absence of gravitational forces. This allows investigation on fluid dynamic effects, phenomena that are normally masked by gravity driven convection, sedimentation, stratification and fluid static pressure. These effects include e.g. diffusion-controlled heat and mass transfer in crystallization processes, interfacial mass exchange, simulation of geophysical fluid flows, emulsion stability and many more.

An Experiment Container (individually-developed for each specific FSL experiment) is inserted into the Central Experiment Module drawer by a crewmember. Each Experiment Container holds the fluid cell assembly (including the process stimuli and control electronics) and may also be equipped with dedicated diagnostics to complement the standard Fluid Science Laboratory diagnostics. The Geoflow experiment (Fig. 2.3 right part) which is undertaking an investigation of importance for astrophysical and geophysical problems, such as global scale flow in the atmosphere and oceans took place in the Fluid Science Laboratory, with the follow-up experiment, Geoflow-2 planned in 2011.

The FSL and experiment control concept allows alternative modes of operation consisting of fully automatic, semi-automatic and fully interactive experiment processing. All these modes may be initiated either by the flight crew or from the ground. The investigators can monitor and adjust the processing of their experiments also from their own User Home Bases. The Facility Responsible

Centre for the Fluid Science Laboratory, the Microgravity Advanced Research and Support Centre (MARS) in Naples, Italy which is supported by the Facility Support Centre E-USOC in Madrid, Spain.

European Drawer Rack – Multi-discipline Experiment Rack in Columbus for Subrack Experiments

The European Drawer Rack, EDR, [30] in Columbus was developed by ESA to provide a flexible facility for accommodating medium-sized dedicated experiment equipment covering a variety of scientific disciplines. The facility can accommodate up to three standard 72 l drawers, or four standard 57 l lockers. This approach allows a quick turn-around capability, and provides increased flight opportunities. Outside of resources management the operating concept of the European Drawer Rack assumes that payloads are largely autonomous and research teams can monitor their experiments from local User Home Bases. The Facility Responsible Centre for the European Drawer Rack is the Erasmus User Support and Operations Centre at ESA's ESTEC Facility in Noordwijk, the Netherlands. The individual payloads it houses partially have different centres responsible for the associated research activities. The Belgian User Support and Operations Centre for example was responsible for the Protein Crystallisation Diagnostics Facility [34], which undertook 3½ successful months of experimentation in the European Drawer Rack in 2009 in order to help establish the conditions under which good zeolite crystals can be grown. An interface drawer was developed to allow the European Drawer Rack to host a European Kubik incubator (see below). In the future EDR will temporarily accommodate the Facility for Adsorption and Surface Tension (FASTER), and the Electro-Magnetic Levitator (details see below).

European Physiology Modules – European Facility for ISS Physiology Research

The European Physiology Modules [33] is equipped with Science Modules to investigate the effects of long-duration spaceflight on the human body. The experiment results will contribute to an increased understanding of terrestrial problems such as the ageing process, osteoporosis, balance disorders, and muscle wastage. The initial configuration of instrument modules for the facility included: the Multi-Electrode Electroencephalogram (MEEMM) which is dedicated to the study of brain activity by measuring both EEG and Evoked Potentials; the Portable Electroencephalogram Module (PORTEEM) which is a portable EEG/polysomnography module for ambulatory and sleep studies; the Cardiovascular Laboratory or Cardiolab, which is a joint French (CNES) / German (DLR) Module supporting cardiovascular research including blood pressure device, ECG or portable Doppler measurement; and equipment for blood, saliva and urine collection. Additional modules that were launched to enhance the capabilities of the European Physiology

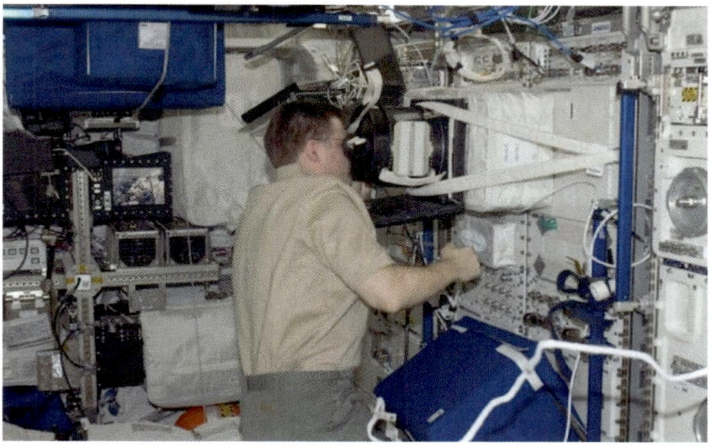

Fig. 2.4 ISS Expedition 25 commander Doug Wheelock (NASA) using Neurospat hardware to perform the PASSAGES experiment with the European Physiology Modules (Source: NASA)

Modules include the Portable Pulmonary Function System (See below). NASA also developed physiology research racks which are complementary to the European Physiology Modules called the Human Research Facility −1 and −2 (located also in Columbus). These offer additional physiological research capabilities such as the assessment of pulmonary function (see Pulmonary Function System below), blood centrifugation, body mass measurement and ultrasound.

The Facility Responsible Centre for the European Physiology Modules rack, CADMOS in Toulouse, France has the overall responsibility to operate it according to the needs of individual Science Modules. PASSAGES (Fig. 2.4), looking into how astronauts interpret visual information is an example of an experiment utilising the capabilities of the European Physiology Modules.

Pulmonary Function System (in Human Research Facility 2)

The Pulmonary Function System [42], which was built by ESA is an ESA/NASA collaboration in the field of respiratory physiology instrumentation, analyses exhaled gas from astronauts' lungs to provide near-instant data on the state of crew health. Used previously for ESA's CARD experiment and for the Periodic Fitness Evaluation it is capable of a wide range of respiratory and cardiovascular measurements. This includes breath-by-breath measurements, diffusing capacity of the lung, expiratory reserve volume, forced expired spirometry, functional residual capacity, cardiac output, alveolar ventilation, volume of pulmonary capillary blood, and other pulmonary tests. For support of European experiments Damec in Odense, Denmark is the responsible User Support and Operations Centre (USOC) for the Pulmonary Function System.

Portable Pulmonary Function System

The Portable Pulmonary Function System which also falls under the operational responsibility of Damec has similar capabilities to the Pulmonary Function System, but is designed for use in all parts of ISS i.e. it's fully portable and not mounted in a rack. This makes it useful for metabolic measurements during exercise, for example. It has been used for undertaking ESA's Thermolab experiment, which investigates thermoregulatory and cardiovascular adaptations during rest and exercise during long-term exposure to weightlessness, in conjunction with NASA's Maximum Volume Oxygen (VO2 Max) experiment.

KUBIK Incubator

KUBIK can function as an incubator or cooler (+6°C to +38°C temperature range). Self contained automatic biological experiments can be performed using power provided by the facility. A centrifuge insert permits 1 g control samples to be run in parallel with the weightless samples. If a centrifuge control is not needed it is possible to interface larger, dedicated experiment hardware with KUBIK via an interface plate. KUBIK incubators can be potentially operated powered in Soyuz spacecraft providing a means of maintaining controlled temperature and performing automatic experiments from a just prior to launch until docking with the Station. It was used in 2010 for processing the PADIAC and SPHINX experiments in the European Drawer Rack.

Facility for Adsorption and Surface Tension Studies (FASTER)

From 2012 FASTER will study the links between the physical chemistry of the droplets interface, the liquid films and the collective properties of an emulsion. A relevant problem in emulsion technology is the control of emulsion stability. For example high stability is necessary for emulsions in foods, cosmetics, pharmacy etc. whereas separation is required in waste water processing and oil recovery.

Electro-Magnetic Levitator (EML)

The ESA/DLR Electro-magnetic Levitator will perform containerless materials processing from 2012, involving melting and solidification of electrically conductive, spherical samples, under ultra-high vacuum and/or high gas purity conditions. Heating and positioning of the sample is achieved by electromagnetic fields generated by a coil system. EML will support research in the field of meta-stable states and phases and in the measurement of high-accuracy thermo-physical properties of liquid metallic alloys at high temperatures up to 2,000°C. The former covers investigations of nucleation and solidification kinetics in under-cooled melts and microstructure formation for instance.

Solution Crystallisation Diagnostics Facility (SCDF)

Biolab – European Rack Facility for Biological Experiments on the ISS

Following the successful completion of the Protein Crystallisation Diagnostics Facility experiments in 2009, the facility will be 'refurbished' as an instrument capable of providing a range of optical techniques of general interest for nucleation, growth and solidification studies within projects dealing with the growth of crystalline and amorphous structures from solutions. The Solution Crystallisation Diagnostics Facility will be hosted in the European Drawer Rack as an advanced light scattering instrument that combines state of the art static and dynamic light scattering, ultra low angle scattering and more recent multi-speckle techniques based on cameras.

European Modular Cultivation System

The European Modular Cultivation System [44] is an ESA gravitational biology payload installed inside a NASA EXPRESS rack in Columbus. The facility is dedicated to experiments on plants, especially multi-generation (seed-to-seed) experiments and studies gravity effects on early development and growth, on signal perception and transduction in plant tropisms. Experiments with insects or amphibia as well as studies with cell and tissue cultures can also be conducted.

The facility consists of a gas tight incubator with two centrifuges and space for four Experiment Containers on each rotor. The rotors contain systems for life support, water supply, illumination and observation. A facility laptop, gas supply module and thermal control system are located outside the incubator. The crew will set up experiments and exchange containers for resupply of consumables such as gas and water. Otherwise, the facility operates autonomously and can be controlled from the ground or by the crew. Ground and on-orbit reference experiments can be undertaken. A Test Bench with EMCS ground models is established at the University of Trondheim, Norway (location of the User Support and Operations Centre for the European Modular Cultivation System) to validate various aspects of experiment development.

Portable Glovebox

The Portable Glovebox has been used for the handling of various biology experiments with the European Modular Cultivation System and the Kubik incubators. It provides an adequate enclosure to perform manual operations during safety critical steps of any experiment. The Glovebox has an airtight volume of 21 l, with two gloves mounted on standard glove rings for handling the experiment equipments inside.

Muscle Atrophy Research and Exercise System (MARES)

MARES [45], is a general-purpose facility for (neuro-)muscular and exercise research. It is capable of assessing the strength of isolated muscle groups around joints by controlling and measuring relationships between position/velocity and torque/force as a function of time. This is done during passive exercises on MARES, while its motor puts a programmed load on the astronauts' extremities movements. It is an ideal tool for research on the countermeasure efficiency and a vast improvement on current muscle research facilities on the ISS. MARES consists of an adjustable chair with a system of pads and levers that fit to each astronaut and a main box containing the facility motor and control electronics. MARES is capable of acquiring data, controlling and providing power to external devices (such as the Percutaneous Electrical Muscle Stimulator, see below) and transferring real time data for downlink. Research teams can monitor the execution of their experiments from local User Home Bases. The Facility Responsible Centre for the European Physiology Modules facility, CADMOS in Toulouse, France has the overall responsibility to operate it according to the needs of individual experiments.

Percutaneous Electrical Muscle Stimulator (PEMS)

PEMS is a second generation of a device which already flew on the Space Shuttle in 1996 and which has been also deployed on ISS as part of EPM/HRF. Its purpose is to deliver electrical charge pulse stimulation to non-thoracic muscle groups of the human test subject, thereby creating contractile responses from the muscles. Its main purpose is to support human neuromuscular research.

Flywheel Exercise Device

The Flywheel Exercise Device (Fig. 2.5) [43] is an advanced exercise device for ISS astronauts and for human physiology investigations in the area of advanced crew countermeasures. Acting to counter muscle atrophy, bone loss, and impairment of muscle function in astronauts during long duration spaceflights, the exercise device uses a rotating flywheel that replaces weight plates and other means of resistance training that rely on gravity. The resistance is provided by spinning flywheels with a cord being wound and unwound around the axle of a fixed flywheel.

Eye Tracking Device

The Eye Tracking Device (ETD), which was developed by DLR, consists consists of a headset that includes two camera modules for binocular recording of horizontal,

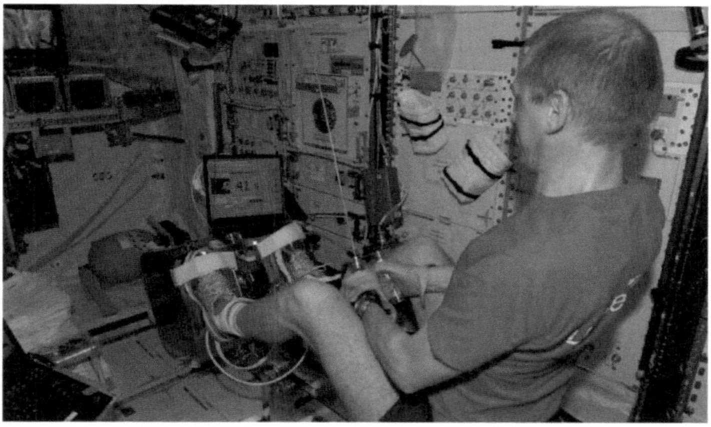

Fig. 2.5 ESA astronaut and ISS Expedition 21 Commander Frank De Winne checking out the ESA Flywheel Exercise Device, with the in the Columbus laboratory of the International Space Station (Source: NASA)

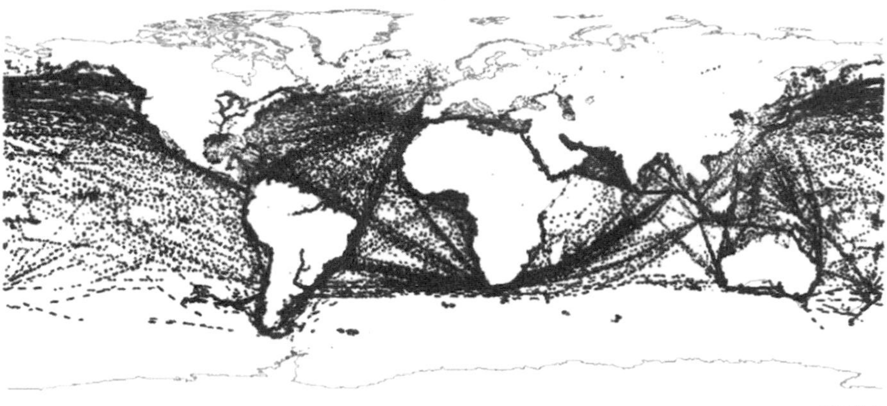

Fig. 2.6 Accumulated plot of AIS position reports received from 2 to 7 June 2010. Ship routes appear clearly (Source: FFI)

vertical and rotational eye movements. The device also includes a laptop to record and process results including continuous images of the subject's eyes. The device was first used as part of the Dutch / ESA DELTA mission in 2004 and has been used subsequently on numerous occasions. Meanwhile ETD has been disposed.

Vessel Identification System

The Vessel Identification System is testing the means to track global maritime traffic from space by picking up signals from standard AIS transponders carried by all international ships over 300 tonnes, cargo vessels over 500 tonnes and all types of passenger carriers (Fig. 2.6). The system consists of two different receivers (NORAIS and LuxAIS), which are alternated every 3 months or so, and the ERNO-Box, which is used as a data relay for the Vessel Identification System, whose antenna is located on the outside of Columbus. Data is received by the Norwegian User Support and Operation Centre (N-USOC) in Trondheim via ESA's Columbus Control Centre in Germany. The success of this technology opens other possibilities for global tracking from space in other areas.

2.2.1.3 Other European Research Facilities on the ISS

Microgravity Science Glovebox

The Microgravity Science Glovebox (Fig. 2.7) [37] was the first European-built research rack facility to be launched to the ISS in 2002. It was developed by ESA within a barter agreement with NASA. The Glovebox provides the ability to perform a wide range of experiments in the fields of material science, biotechnology, fluid science, combustion science and crystal growth research, in a fully sealed and controlled environment. The facility was relocated from Columbus back to the US laboratory in October 2010. The 'gloves' are the access points through which astronauts can manipulate experiments. The facility can maintain an inert atmosphere with dry nitrogen and less than 10% oxygen and offers many different command and control capabilities to allow performance of investigations attended or not by the crew. The Core Facility occupies the upper half of the overall rack and

Fig. 2.7 ESA astronaut Frank De Winne during installation of the selectable optical diagnostic instrument hardware into the Microgravity Science Glovebox in 2009 (Source: NASA)

includes the large sealed working volume, an airlock, and electronics for control, housekeeping and investigation resources. The Command and Monitoring Panel monitors the facility status and performance and provides all means for the manual operation of the facility by the crew. Numerous experiments have taken place in the facility, an example of which is the SODI series of experiments (See below).

- SODI Instrument and Experiments
  The Selectable Optical Diagnostic Instrument (SODI) combines different optical techniques in a single instrument. It is equipped with two Mach-Zehnder interferometers that can be operated at two wavelengths and allow for scanning of multiple cells in a cell array as well as a piezo-activated mirror which allows stepping for phase determination. SODI can also be equipped with a Near Field Scattering instrument. The experiments to date have covered the influence of vibrations on diffusion in liquids (SODI-IVIDIL) and the study on growth and properties of advanced photonic materials within colloidal solutions (SODI-Colloid).
- Directional Solidification (DIRSOL) experiment
  The DIRSOL facility (under development) aims to perform DIRectional SOLidification experiments of transparent materials using the Bridgman technique and will complement the capabilities of DECLIC (See below). DIRSOL will be available from the end of 2011 and will be positioned in the Microgravity Science Glovebox. The main diagnostics element of DIRSOL is optical observation with high resolution. The observation camera can observe the transparent samples between the hot and cold zones of the Bridgman assembly at variable positions and viewing angles.

Material Science Laboratory in the Material Science Research Rack

The Material Science Laboratory [36] is the primary research facility located in NASA's Materials Science Research Rack-1, which was launched on STS-128/17A, together with a total of six sample cartridges for NASA and for ESA's MICAST and CETSOL projects under a cooperation agreement with NASA and is now installed in the US Laboratory on the ISS. The Core Facility of the Material Science Laboratory is a sealed stainless steel cylinder (the Process Chamber) capable of accommodating different individual furnace inserts, within which sample processing is carried out. Processing conditions are normally either a vacuum or an inert gas (*e.g.*, Argon).The different inserts are the Low Gradient Furnace and the Solidification and Quenching Furnace, which are based on the Bridgman Technique with a hot and a cold zone and an adiabatic zone in between. The crew insert an experiment cartridge (holding a sample to be processed) into the furnace (Fig. 2.8 left). Following process chamber evacuation, an experiment sequence is initiated, consisting of a number of steps with pre-defined parameters. Though experiment execution is automatic, processing parameters can be altered from ground.

Fig. 2.8 *Left*: ESA astronaut and ISS Expedition 21 Commander Frank De Winne, exchanging samples of the CETSOL/MICAST experiments in November 2009. *Right*: ESA astronaut and Expedition 14 Fight Engineer Thomas Reiter working with the European Matroshka Phantom on the International Space Station (Source: NASA)

The Facility Responsible Centre for the Material Science Laboratory, the Microgravity User Support Centre (MUSC) in Cologne, Germany, has the overall responsibility to operate it according to the needs of individual research objectives.

DECLIC

The DEvice for the study of Critical LIquids and Crystallization (DECLIC) is an apparatus developed by CNES for a NASA EXPRESS rack to support the study of material growth and liquids behaviour near their critical point. It provides all subsystems required to operate an experiment dedicated insert installed on an optical bench. To date three experiment modules have been developed for DECLIC: the former ALICE Like Insert, dedicated to the observation of gas-liquid transformation near the critical point of pure fluids at low (near-room) temperatures; the High Temperature Insert for heat and mass transfer at high temperature and pressure, and the Directional Solidification Insert for observation of microstructures that form at the liquid-solid interface when transparent material solidifies. The instrument will allow observing the interaction of supercritical water in presence of a liquid or solid solute (dissolution, mixing, jet injection, cold combustion, and nucleation) through imaging and interferometric optical techniques.

Radiation Monitoring Facilities and Devices

It is of key importance to monitor and understand the cosmic radiation levels that are experienced by astronauts in orbit. The dosimetric facilities and equipment are key to building our fundamental knowledge of the radiation environment and help to test new types of shielding material for use on future human spaceflight missions and possibly for applications on Earth. In addition to various radiation monitoring devices developed by the non-European ISS partners, Europe has also developed a suite of instruments and facilities related to radiation research:

- Matroshka
  Matroshka [38] is a special ESA payload that has been in use on the ISS since 2004 to determine radiation levels affecting astronauts at different depths in the human body either inside the ISS or during spacewalks. Matroshka experiments consist of a simulated human body (head and torso) called the Phantom (Fig. 2.8 right) which simulates the human body with relation to its size, shape, position, mass, density and nuclear interactions. It is composed of natural bone and a material resembling natural tissue, built up in layers. A lower-density material simulates the lungs. The Phantom is covered with a simulated skin layer and can be housed in an external container, to represents a spacesuit. The facility is equipped with several active and hundreds of passive dosimeters located at various positions to measure doses of all different types of radiation at key organ sites. The active dosimeters provide time-dependent measurements, whilst the passive dosimeters are analysed after flight to determine accumulated doses. During its time on the ISS the facility has undertaken extensive measurement campaigns both outside and insidevarious ISS modules.
- Alteino/SilEye-3
  This active cosmic ray detector developed nationally in Italy is built upon research with two previous SilEye cosmic ray detectors and associated research that was undertaken on the Mir Space Station from 1995 to 1999. The Alteino device (which consisted of the SilEye-3 detector and an Electroencephalograph) was flown to the ISS as part of the ESA/ASI Marco Polo mission in 2002 and was central to an experiment to test the effects of radiation on the electrical activity of the brain. Subsequently the Alteino device played a central role in additional experiments such as ALTCRISS (ALTEINO Long Term monitoring of Cosmic Rays on the ISS).
- ALTEA
  The follow-up to the Alteino device was the Anomalous Long Term Effects in Astronauts' Central Nervous System (ALTEA) device (Fig. 2.9) which consisted of a helmet-shaped structure consisting of six silicon particle detectors and an EEG to measure brain activity. The hardware was developed by ASI and has been used in a number of previous experiments including ESA's ALTEA-Shield experiment, which aims to improve understanding of the light flash phenomenon, tests the effectiveness of different types of shielding material, and is undertaking a 3-dimensional survey of the radiation environment in the ISS.

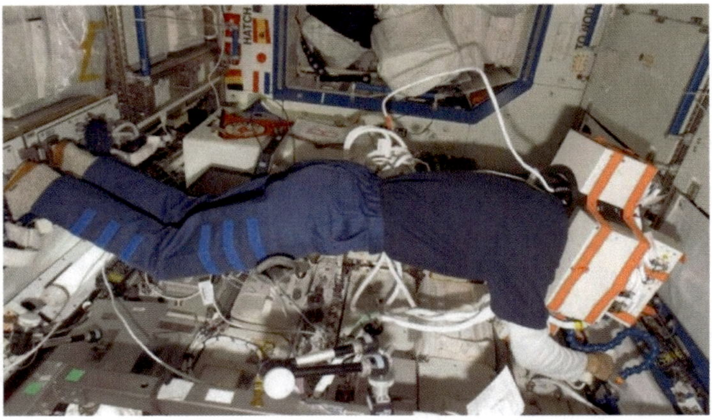

Fig. 2.9 NASA astronaut and Expeditions 14 and 15 Flight Engineer, Sunita Williams, wears the anomalous long term effects in astronauts' central nervous system (ALTEA) experiment helmet while conducting the experiment in the US laboratory (Source: NASA)

In the Columbus laboratory itself two active **DOSTEL radiation dosimeters** are located inside the European Physiology Modules facility for taking part in experiments such as the Dose Distribution inside the ISS (DOSIS) experiment.

Plasma Kristall-4 and -3 Plus (PK-4 and PK-3 Plus) Payloads

The PK-4 payload will perform novel research in the field of 'Complex Dusty Plasmas'. These are low-temperature gaseous mixtures composed of ionized gas, neutral gas and micron-sized particles. The micro-particles become highly charged in the plasma and interact strongly with each other through the Coulomb force. These interactions can, in specific conditions, lead to a self-organized structure of the micro-particles: so-called plasma crystals. PK-4 will consist of a glass-made DC discharge plasma chamber. The elongated DC-plasma chamber of PK-4 is especially suited for investigations of complex plasmas in the fluid phase. On ground such experiments are distorted by gravity. This builds on previous complex plasma research in weightlessness which has been ongoing since 1998 and includes the ISS PK-3 plus experiment (and hardware) which is still ongoing and was developed by DLR.

Cardiomed

Cardiomed is CNES supplied and consists of Cardiolab type instruments which measure parameters such as muscular activity of the myocardium (electrocardiogram), blood pressure and arterial blood flow, being monitored in real time from the Mission Control Centre in Moscow (MCC-M). Data gathered will identify mechanisms related to the effects of weightlessness on the cardiovascular system. This can help in the development of countermeasures to keep the crew in good health.

Elaborator of Televised Images in Space (ELITE-S2)

ELITE-S2 is an ASI system for observations on body motor control during long term exposure to microgravity and to perform quantitative data collection and analysis on board the ISS. Launched in 2007, ELITE-S2 studies the strategies for dynamic control of posture and body motion and adaptive mechanisms which allow adjustment of motor control strategies resulting from exposure to microgravity. It can allow investigations on the effects of weightlessness on breathing mechanisms, studies on the adaptive mechanisms which allow dynamic adjustment of motor control and posture control strategies resulting from exposure to microgravity, and applications of ergonomics findings in the design of spacecraft.

Hand Grip Dynamometer/Pinch Force Dynamometer

The Handgrip Dynamometer is a handheld device capable of measuring instantaneous hand strength as a function of time. The principal components are a handgrip, instrumentation amplifier and cables. The Pinch Force Dynamometer is a handheld device capable of measuring instantaneous strength of the thumb and opposing finger or groupings of fingers as a function of time. The principal components are a pinch force transducer, instrumentation amplifier and cables.

Mice Drawer System

The ASI Mice Drawer System is a general purpose compact animal research facility mainly for use with small rodents (mice). It can provide the habitats for six mice for up to 100 days (with possible extension up to 180 days). During its first flight it supported research into human bone formation and osteoporosis prevention countermeasures.

Minus Eighty Degrees Laboratory Freezer for the ISS (MELFI)

The frozen and refrigerated capabilities on the ISS are very extensive and of high importance namely for life sciences. MELFI (Fig. 2.10) [39] is the principal facility, developed in Europe, offering these capabilities on the ISS. Currently there are three European-built MELFI freezers on the ISS: two in the Japanese laboratory and one in the US laboratory. The MELFI freezers are rack-sized units

Fig. 2.10 ESA astronaut and ISS Engineer Paolo Nespoli, services the Minus Eighty *deg* Laboratory Freezer for ISS (MELFI-1) in the ISS Kibo laboratory in December 2010 (Source: NASA)

with four individual dewars that can be used for storage or fast freezing physiological and biological samples with a total capacity of 300 l. Each vacuum-insulated dewar can independently be set to one of the three different operating temperatures (−80, −26 and +4°C).

2.2.1.4 European Space Exposure Research Outside the ISS

Columbus External Payload Facility

The Columbus laboratory is also fitted with an external facility for attachment of research equipment requiring exposure to the open space environment. This has four locations: one pointing towards the Earth (NADIR), one pointing away from the Earth (ZENITH) and two pointing in a starboard direction to the direction of flight (LIMB) of the ISS. Payloads are integrated onto Columbus External Payload Adapters (CEPA) prior to launch to the Station. Once installed by EVA or robotic means payloads can be automatically supplied with power and data connections.

External Facilities

When Columbus was first installed on the ISS it was initially outfitted with the European Technology Exposure Facility (EuTEF) [40], which carried a suite of nine different payloads with 13 different experiments in materials research, space physics, astrobiology, radiation, and space technology.

- DEBIE-2: 'DEBris In orbit Evaluator'
- MEDET: Materials Exposure and Degradation ExperimenT orbit.
- TRIBOLAB: Experiments for research in tribology.
- FIPEX: Sensor for measurement of atomic oxygen
- PLEGPAY: PLasma Electron Gun PAYload
- DOSTEL: DOSimetric radiation TELescope
- EXPOSE: Five exobiology experiments
- EuTEMP: Multi-input thermometer
- EVC: Earth Viewing Camera

SOLAR Facility

The SOLAR payload facility (Fig. 2.11) [35] has been studying the Sun's irradiation with unprecedented accuracy across most of its spectral range since 2008, producing excellent scientific data during a series of Sun observation cycles. It will continue to gather science data in a period of increasing solar activity up to 2013 and possibly beyond. SOLAR consists of three instruments complementing each other to allow measurement of the solar flux throughout virtually the whole electromagnetic spectrum – from 17 to 3,000 nm – in which 99% of the solar energy is emitted. The instruments are mounted on a device for accurate Sun pointing and are controlled by a Control Unit. The scientific instruments are: SOVIM (SOlar Variable & Irradiance Monitor) which covers near UV, visible and thermal regimes (200 nm–100 μm) SOLSPEC (SOLar SPECctral Irradiance measurements) which covers the 180–3,000 nm range with high spectral resolution and SOL-ACES (SOLar Auto-Calibrating EUV/UV Spectrophotometers) which measures the EUV/UV spectral regime (17−220 nm) with moderate spectral resolution.

Fig. 2.11 The solar payload facility pictured on the Columbus laboratory during the STS-124 mission (Source: NASA)

EXPOSE-R

The Expose-R facility hosts a suite of nine astrobiology experiments (eight from ESA, one from the Russian Institute for Biomedical Problems (IBMP), in Moscow), some of which could help understand how life originated on Earth. The facility accommodates experiments in three special sample trays, which are loaded with a variety of biological samples including plant seeds and spores of bacteria, fungi and ferns, which are exposed to the harsh space environment (Solar UV, cosmic radiation, vacuum and microgravity), for about 2 years.

In the future the Columbus External Payload Facility will accommodate:

Atomic Clock Ensemble in Space (ACES)

ACES [41] accommodates two atomic clocks: PHARAO ('Projet d'Horloge Atomique par Refroidissement d'Atomes en Orbit') is a primary frequency standard based on samples of laser cooled cesium atoms developed by CNES; SHM (Space H-Maser) is an active hydrogen maser for space applications developed by ESA. The functionality and performance of this new generation of atomic clocks will be tested on the ISS for at least two years. Comparisons between distant clocks both space-to-ground and ground-to-ground will be performed worldwide with unprecedented resolution. These comparisons will be used to perform precision tests of the special and general theory of relativity. ASIM has national co-funding from Denmark, Norway, and Spain.

Atmospheric Space Interactions Monitoring Instrument (ASIM)

ASIM will study giant electrical discharges (lightning) in the high-altitude atmosphere above thunderstorms and their role in Earth's climate. The external payload is composed of light detectors, sensitive in the optical range (cameras, photometers) and in the X-ray to Gamma-ray ranges (imaging spectrometer).

2.2.1.5 European Control Centre Network

To harness the diverse European involvement in the International Space Station, Europe through the European Space Agency defined a network of ground control centres across the continent to control Europe's infrastructure on the ISS [22].

Columbus Control Centre

The Columbus Control Centre (Fig. 2.12) is located at the German Space Operations Centre of the German Aerospace Center (DLR), in Oberpfaffenhofen,

Fig. 2.12 Overview of room K4 at the Columbus control centre in Oberpfaffenhofen, Germany (Source: DLR)

near Munich, Germany. Operational 24 h a day, 7 days a week, the Columbus Control Centre operates Europe's Columbus laboratory on the ISS with responsibility for coordinating all the Columbus system and research activities. It is in close contact with the Mission Control Centre in Houston, USA, which has overall responsibility for the ISS, together with the Mission Control Centre in Moscow, Russia. In addition, the Columbus Control Centre coordinates operations with the ISS Payload Operations and Integration Centre at the Marshall Space Flight Centre in Huntsville, Alabama, USA, which is responsible for the overall US utilisation activities namely in the Destiny laboratory and other partner laboratories.

The Columbus Control Centre provides the ground services for Columbus operations including communication services (voice, video and data) to all sites responsible for control and coordination of individual European facilities and experiments on the ISS, including the network of User Support and Operations Centres across Europe, industrial support sites, home bases of research teams, as well as ESA management and associated ESA locations.

User Support and Operations Centres (USOCs)

USOCs are based in national centres throughout Europe and are responsible for the use and operation of European payloads on board the ISS. On behalf of users (i.e. scientists, technology developers, education specialists etc.) and under ESA contract, the USOCs conduct the tasks needed to prepare and operate the research facilities and experiments on board the ISS. The USOCs act as the link between the user and their ISS research, and are the focal point for the preparation and operation of ESA payloads on the ISS. All European USOCs are centrally linked to the ISS via the Columbus Control Centre or control centres of ISS partner agencies for ESA payloads located outside of Columbus. There are three levels of responsibility for USOCs. The first is as a Facility Responsible Centre which has the responsibility for a specific multi-user rack level facility such as the European Drawer Rack for example and includes assisting scientists with payload operations. The next level of responsibility is the Facility Support Centre that is responsible for a subrack payload in a rack facility such as the Electro-Magnetic Levitator that will fly in the future and be located within the European Drawer Rack. The last level of responsibility is the Experiment Support Centre which has the responsibility for single experiments either as self-standing experiments or within a facility, and mainly focuses on science and experiment operational matters. Dedicated User Home Bases can also be set up where scientists can carry out real-time data monitoring and control of their respective experiments.

The current network of User Support and Operations Centres includes the Belgian User Support and Operation Centre (B.USOC) in Brussels, Belgium; the Biotechnology Space Support Centre (BIOTESC) in Zurich, Switzerland; the Centre d'Aide au Développement des activités en Micro-pesanteur et des Opérations Spatiales (CADMOS) in Toulouse, France; Damec Research Aps (DAMEC) in Odense Denmark; the Erasmus User Support and Operations Centre in Noordwijk, the Netherlands; the Spanish User Support and Operations Centre (E-USOC) in Madrid, Spain; the Microgravity Advanced Research and Support Centre (MARS) in Naples, Italy; the Microgravity User Support Centre (MUSC) in Cologne, Germany, and the Norwegian User Support and Operations Centre (N-USOC) in Trondheim, Norway. In addition to the research-related control centres mentioned above, the European Space Agency also has an ATV Control Centre based on the premises of the French space agency, CNES, in Toulouse, France, which is responsible for operating Europe's Automated Transfer Vehicle, Europe's logistics supply spacecraft for the ISS.

2.2.2 Foton/Bion Recoverable Capsules

Europe has been an extensive user of the Russian Foton, Bion and Resurs-F type of recoverable unmanned capsules [17] for carrying out research in low Earth orbit since 1975. They were based on the Vostok spacecraft, which carried Gagarin as the first man into space in 1961 and the Zenit military reconnaissance satellite. A typical Foton mission lasts about 12–18 days and as the mission is unmanned all dedicated experiments have to be fully automated with telemetry allowing for up and downlink capabilities for command and control of payloads and experiment parameters. After being put into orbit by a Soyuz-U launcher from the Baikonur Cosmodrome in Kazakhstan (previously the Plesetsk Cosmodrome in Russia) the spacecraft places itself in the correct orientation using attitude control thrusters, before undertaking a free- flying period which will last until the final day of the mission when the re-entry procedures start. The up to 650 kg of research and support equipment for experiments are housed in the spacecraft's Re-entry Module

GradFlex	(ESA-*55kg*)	**2 fluid physics experiments**
TeleSupport	(ESA-*24kg*)	**assists all payloads on board**
Biopan	(ESA-*27kg*)	**10 experiments in exobiology and radiation exposure**
SCCO	(ESA/CSA-*28kg*)	**3 experiments on diffusion effects in crude oil**
Biobox	(ESA-*67kg*)	**5 experiments on cellular biology**
Eristo/Osteo	(CSA/ESA-*71kg*)	**6 experiments on bone growth and yield**
AquaHab	(DLR/ESA-*18kg*)	**2 experiments in biology of water organisms**
Polizon	(KBOM-*182kg*)*	**7 cooperative experiments on material science**
Stone	(ESA-*1kg*)	**2 meteoritic re-entry experiment**
Granada	(ESA-*5kg*)	**growth of several protein crystals (2 exp.)**
Freqbone	(B/ESA-*12kg)*	**countermeasures for bone losses in μg (1 exp.)**
YES-2	(ESA-*40kg*)	**student payload, tether-assisted re-entry demonstrator**
DataLogger	(ESA/TsSKB-*2kg*)*	**measurement of shocks, temperature, and RH in Foton**
Dimac	(ESA-*9 kg*)*	**tri-axial accelerometer system (true DC to 200Hz)**
Battery	(ESA-*15kg*)*	**Li-ion primary batteries for re-entry and landing**
Teplo	(B/ESA-*17kg*)**	**low-g performances of new design (loop) heat pipes**
OWLS	(ESA-*0.5kg*)*	**wireless technology demonstrator** ***(part of TeleSupport)***
SEEK	(ESA-*0.2kg*)*	**measurement of g-loads** ***(part of Gradfiex)***
Photo-II	*(ASI/ESA-4kg)***	**space radiation effects on photosynthesis**
Total	***348 (394) kg***	***(*) not contractually accounted (**) nationally covered flight***

Fig. 2.13 European Scientific payload on Foton-M3

(or outside the module in the case of astrobiology/exposure experiments). The spacecraft also includes a Service Module responsible for attitude control and deorbit retroboost of the whole spacecraft and a Battery Module supplying electrical power to the spacecraft and research equipment. Prior to re-entry the three modules separate and the Re-entry Module, which is covered in an ablative material, re-enters the Earth's atmosphere, undertaking a parachute-assisted landing where it is retrieved. The other two modules burn up in the atmosphere as planned.

Foton was envisaged as a platform for physics and materials science to complement the Bion capsules that were aimed at life science studies. However, in later years increasing numbers of experiments were transferred to Foton, while the Bion programme was temporarily discontinued in 1996. Scientific payloads from western Europe have regularly flown on Foton. On the flights of Foton-M2 and Foton-M3 for example there was an extensive amount of research equipment totalling 660 kg (Fig. 2.13) across both flights. This included experiments such as Gradflex which was looking into density/concentration fluctuations in fluids exposed to thermal gradients in weightlessness; and exposure experiments in the Biopan facility which was trying to answer fundamental questions about the origin and spreading of life forms in space.

2.3 Non-orbital Weightless Research Platforms

Outside of the orbital infrastructure discussed in the previous section, Europe (and other space-faring nations) make use of additional research platforms that provide access to varying degrees of weightlessness. These can either be a sufficient end point for research goals or may act as a precursor to flying experiments and hardware in orbit. The principal platforms for this area of research are:

2.3.1 *Sounding Rockets*

Originally conceived to sound the physical properties of the upper atmosphere, hence the name 'sounding' rockets, their use since the late 1950s has been extended beyond Meteorological and Upper Atmosphere studies to provide 'weightless' conditions for experimental research in physical and life sciences. Sounding rockets [16] are a sort of ballistic missile able to launch a few hundred kilograms to altitudes of 250–750 km, with almost vertical ascent and descent trajectories.

Excellent weightless conditions are met during the freefall phase of the rocket's payload once the rocket motors have exhausted their thrust and separated from the payload. The rocket's payload continues to rise under the momentum built up during the launch phase before falling back towards Earth. The period of weightlessness ends just prior to its re-entry phase. The freefall ends with the re-entry in the Earth atmosphere and finally the deployment of the parachute that lowers the payload to the ground with appropriate impact speeds (Fig. 2.14 right).

This microgravity research platform provides a great flexibility as there are a variety of different sounding rockets available to be utilised by the European scientific, industrial, commercial and education communities. This makes them excellent candidates for performing a experiments with different requirements as well as providing a means of testing experiments and equipment to be undertaken or used in orbit. Sounding rocket missions are a valuable means to undertake

Fig. 2.14 (*Left*) the XRMON-Diff module (prior to launch) which flew on MAXUS 8 in March 2010. (*Right*) MAXUS 7 payload following landing on 2 May 2006 (Source: SSC)

experiments having stringent pre-/post-flight experiment logistics and may use hazardous materials or equipment (e.g. radiation sources for in-situ visualisation). They are also advantageous for high-temperature materials experiments as the heater/furnace materials degrade quickly and as such do not make them candidates for long-term ISS experiments. The duration of weightlessness provided by different sounding rockets ranges from 6 to 12 min with a payload mass of up to 800 kg.

From a European perspective a majority of sounding rocket launches are undertaken at the Esrange launch site in Northern Sweden. It is owned and managed by the Swedish Space Corporation (SSC) with ESA as a regular user and major research partner along with other national organisations and partners such as DLR. The location, just within the Arctic Circle, is excellent for the investigation of polar phenomena. It has a safe impact area of 9,000 km^2; and the launch site has excellent transport links. Sounding rocket services are also offered by the Andoya Rocket Range in Norway. Both Esrange and Andoya have a full range of capabilities at their launch sites for testing and preparation of experiment payloads (Fig. 2.14 left) prior to launch and monitoring them in flight. The availability of real-time data allows experimenters to follow and if necessary direct the course of their experiments. Launch preparation activities allow for a late access to the experiment modules.

Besides the experiment modules other functional modules complement each sounding rocket mission, such as the service module for telemetries and telecommands as well for the rocket attitude control, the video module to relay multiple video channels to the receiving ground stations, the recovery system that commands the deployment of the parachute packet in the nosecone of the rockets, and the separation module needed to detach from the payload stack when the rocket motors are at the end of their thrust (Fig. 2.14 left).

Sounding Rockets that are and have been central to European weightless research activities include TEXUS which has been launched 48 times since 1977 the last launch being TEXUS-47 in November 2009; MASER which has been launched 11 times since 1987, the last launch being MASER-11 in May 2008; and MAXUS which has been launched nine times since 1991, the last occasion being MAXUS 8 in March 2010. At the time of compilation the launch of MASER-12 is scheduled for November 2011 and the launch of TEXUS-49 was scheduled for March 2011 (with TEXUS-48 following in November 2011). The launch of MAXUS-9 is planned for 2013 with an experiment complement of four modules in materials research and biology. The REXUS rocket programme is also a central component of educational activities (together with the BEXUS high-altitude balloon programme) having flown eight times in total. The launches of REXUS 9 and 10 are scheduled for March 2011.

2.3.2 *Parabolic Flight Airplanes*

Parabolic flights [15] are used to conduct short-term scientific and technological investigations in weightlessness and reduced gravity, to test instrumentation prior to

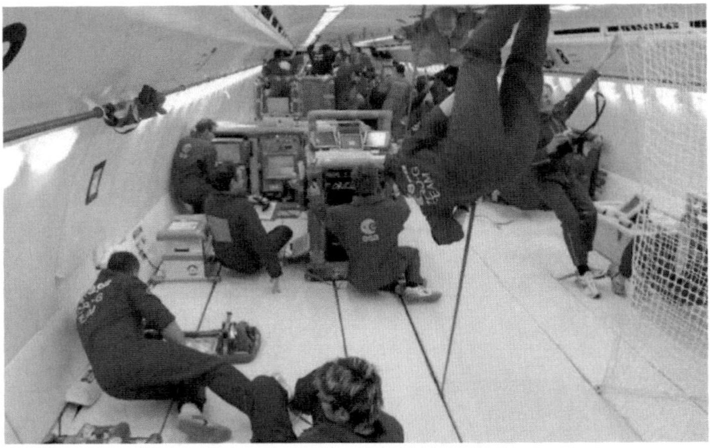

Fig. 2.15 A view inside the 'Zero G' cabin during the weightless phase of a parabola as part of the 49th ESA Parabolic Flight Campaign (Source: ESA – A. Le Floc'h)

use in space, to validate operational and experimental procedures, and to train astronauts for a future human spaceflight missions. In Europe this service is supplied by NoveSpace (a subsidiary of CNES) in Bordeaux, France using an Airbus A300 0-g aircraft with ESA, CNES, DLR all using their services. Such flights are conducted on specially-configured aircraft, and provide repetitive periods of weightlessness. During a campaign, which normally consists of three individual flights, some 30 parabolas are flown on each flight, around 90 parabolas in total. On each parabola, there is a period of increased gravity (1.8–2 g) which lasts for about 20 s immediately prior to and following a 20 s period of weightlessness (Fig. 2.15).

The Airbus A300 zero-g aircraft is now also certified for flying reduced gravity parabolas of 0.16 g for approximately 23 s and 0.38 g for approximately 30 s, which correspond to Lunar and Martian gravity levels. In June 2011 ESA undertook jointly with CNES and DLR the first partial gravity parabolic flight campaign in preparation for future exploration activities. Parabolic flights are the only sub-orbital carrier to provide the opportunity for the science community to carry out medical and physiological experiments on human subjects under conditions of weightlessness or reduced gravity, complementing studies conducted in space, and in simulated conditions on ground (*e.g.* immersion, bed-rest). They also provide physicists with the opportunity of carrying out hands-on investigations on processes characterised by short time scales. These flights offer a flexible approach and short lead-times for researchers, as well as the opportunity to modify their experiments during a flight campaign. Organisations, such as ESA for example, also cover the cost of such flights for research proposals that are selected.

ESA and CNES have had a close partnership on parabolic flights from 1988 when CNES made their Caravelle zero-g aircraft available to ESA for parabolic flight campaigns. This accounted for 15 campaigns until 1995. To date ESA have undertaken more than 54 parabolic flight campaigns covering research and education. CNES have also undertaken around 50 parabolic flight campaigns. The German Aerospace Agency (DLR) have undertaken almost 20 campaigns on the A300 aircraft since 1999 (which included the 10,000th parabola by the A300 0-g aircraft).

2.3.3 *Drop Towers*

Drop Towers [14] offer the opportunity to undertake a variety of experiments in fields such as fluid dynamics, process engineering, combustion, material science, biology and biotechnology requiring only a limited exposure to weightlessness to the extent of a few seconds either due to this suiting experiment requirements or as a precursor to the experiment being flown on a different weightless platform. In Europe the principal exponent of this is at the University of Bremen in Germany within the Center of Applied Space Technology and Microgravity (ZARM) which celebrated its 25 anniversary (and 20th of the Drop Tower) in 2010.

The ZARM Institute houses a variety of experiment facilities and laboratories that are available for ZARM scientists but also for researchers from other universities and organizations. Among these facilities are vibration and aerodynamics test labs as well as facilities that allow performing experiments under micro or hyper gravity conditions. The main laboratory is the 146 m drop tower which, in comparison to orbital systems, represents an economic alternative with permanent access. It serves as an important supplement to either existing or planned orbital or

Fig. 2.16 Internal view of the ZARM drop tower in Bremen, Germany. The drop tower capsule being prepared (Source: ZARM)

suborbital platforms for microgravity research. As of October 2008, 5,000 drop experiments had successfully been carried out since it started operation in 1990.

The installation delivers 4.74 s of near-weightlessness for a single drop up to three times a day. In order to double the microgravity time to 9.3 s, a catapult system was implemented into the drop tower operation routine. With the drop tower catapult, the capsule (Fig. 2.16) performs a quasi-vertical parabola instead of being dropped. Experiments are held in a cylindrical 80 cm wide capsule which is closed pressure-tight. For a drop the capsule is winched up to a height of 120 m. The 1,700 m^3 tube and the deceleration chamber are then evacuated to eliminate aerodynamic forces on the falling capsule. The capsule is then released. A deceleration unit, filled with polystyrene pellets, decelerates the vehicle. For retrieval, the vacuum chamber is reflooded with preconditioned air within 20 min. The experiment and results are hereafter immediately at the scientists' disposal.

At the National Institute for Aerospace Technology in Madrid, Spain a 21 m drop tower, which offers 2.1 s of weightlessness is also available. Experiment payloads can either be placed in a double capsule (for smaller experiments) which offers increased quality weightlessness as the outer capsule acts as a drag shield, or placed in a single capsule for accommodating comparatively larger experiments.

2.3.4 High-Altitude Balloons

High-altitude balloons have been used for decades and the improvement in their capabilities has increased their reliability and potential for undertaking different forms of research such as Ozone studies, circum polar flights, astronomical studies, astrophysical studies, microgravity studies and educational purposes etc. Many different agencies/countries have their own national programmes offering access to high-altitude balloons such as CNES in France, ASI in Italy, DLR in Germany, SSC in Sweden and the Andoya Rocket Range in Norway. As with sounding rockets, two examples of launch sites for high-altitude balloons are the Esrange launch site in Northern Sweden and the Andoya Rocket Range's launch site on the Norwegian Svalbard archipelago.

Many different balloons exist for offering different conditions to the payload that it is carrying with relation to different factors/parameters required such as altitude ranging from ~15–45 km, data measurement time required, and payload mass flown which can vary from a couple of kg to a few tonnes. Launch sites have the necessary facilities to prepare payloads prior to launch, and monitoring payload telemetry. Most balloon systems are used for communication with different service systems or for data transmission from the payload to the ground station.

One example of the use of high-altitude balloons is in the BEXUS (Balloon-borne EXperiments for University Students) programme, which is a collaboration between the Swedish National Space Board and DLR providing students the possibility to undertake their experiments on high-altitude balloons (or on sounding rockets as part of the REXUS Programme). The Swedish share of the payload has been made available to students from other European countries through a collaboration with the European Space Agency.

A typical balloon configuration could consist of a helium-filled balloon, a gondola holding experimental payloads, a cutter to cut the gondola away from the balloon, a parachute system and systems covering altitude control, flight termination, Global Positioning System (GPS), temperature and air pressure sensors, systems for experiment data and control and telemetry. The gondola would typically be retrieved by a helicopter after landing. The last flight of BEXUS was BEXUS 11 which was launched in November 2010. The next scheduled campaign is for BEXUS 12 and 13 in October 2011.

2.4 Ground-Based Facilities and Space Simulators

2.4.1 Bed Rest, Hypokinesia and Metabolic Balance Facilities

Bed rest studies are an invaluable method for simulating the effects of spaceflight on the human body. The physiological effect of lying in bed tilted at 6° for an extended period with the head lower than the feet produces bone/muscle mass loss and fluid shifts similar to those seen in human spaceflight. As such these studies are a cost-effective means of undertaking an analysis of the underlying mechanisms causing these effects and testing different countermeasures (Fig. 2.17) to alleviate the negative physiological effects such as nutritional supplements, exercise

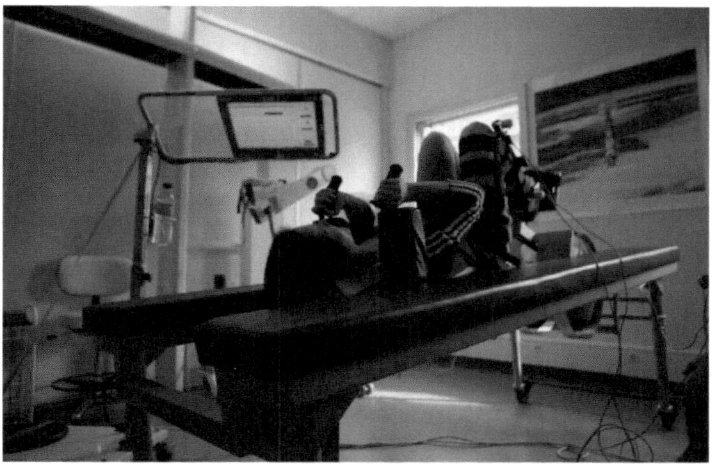

Fig. 2.17 Flywheel exercise device being used as part of experiment protocol during WISE 2005 bed rest study (Source: CNES/Stéphane LEVIN)

protocols and equipment or artificial gravity [27, 28, 50, 51, 71]. The confinement of bed rest subjects also holds similarities with the confinement experienced by astronauts on mission and thus makes these studies helpful in testing psychological protocols.

The **Medes Clinical Research Facility, in Toulouse, France** conducts clinical research studies in a unique European research facility housed within the Rangueil university hospital in Toulouse, mainly in the areas of physiology, pharmacology and the evaluation of biomedical devices. It undertakes experiments which simulate the effects of the space environment (involving bed rest, confinement, circadian rhythms, *etc*) in order to study its physiological effects and to develop preventive methods. A similar facility to Medes is the **Simulation Facility for Occupational Medicine Research (AMSAN), at the DLR Institute of Aerospace Medicine in Cologne**. This multi purpose research facility undertakes bed rest studies with respect to evaluation of protocols/countermeasures though the primary focus of AMSAN is in nutritional studies.

2.4.2 Isolation and Confinement, Pressure Chambers and Climate Chambers

Based on experience gained from isolation studies, combined with other terrestrial-based simulation facilities and data from human spaceflight missions in low Earth orbit, especially within the last two decades, the European scientific and technology community has gained substantial experience in assessing the risks for humans in the space environment. In addition to being helpful in determining psychological aspects of space flight such as the psychology of group dynamics and individual performance under isolation and confinement, this research combined with knowledge obtained from human spaceflight missions has been invaluable in determining human adaptation to conditions in space, as well as in the development of life support systems. Previous space-related isolation studies have included the ISEMSI study in Norway in 1990, the EXEMSI study in 1992 in Cologne Germany, both performed in shore-based hyperbaric chambers, the Human Behaviour in Extended Spaceflight (HUBES) study which modelled aspects of the long-duration Euromir 95 mission and the 'Simulation of the Flight of the International Crew on Space Station' (SFINCSS) from 1999 to 2000.

The **Mars 500 Isolation Facility, in Moscow, Russia** facility is a Russian Institute for Biomedical Problems (IBMP) facility though many IBMP studies have European involvement, one of the most prominent being MARS 500. A purpose built isolation facility was outfitted in order to simulate a human spaceflight mission to Mars (Fig. 2.18). The Mars500 isolation facility in which the crew is based is located in a special building which comprises the isolation facility itself, as well as the operations room, technical facilities and offices. The isolation facility comprises one external module, which is used to simulate the 'Martian surface' and

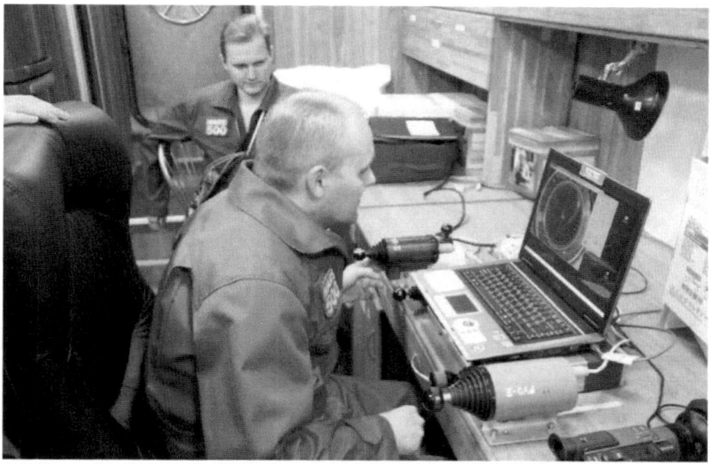

Fig. 2.18 Mars500 crew has a computer simulator for practicing docking, undocking and maneuvering their spaceship. It was also used when ship 'left' Earth orbit for interplanetary cruise on 14 June (Source: ESA)

four hermetically sealed interconnected habitat modules which simulate the spacecraft that take the crew on the simulated journey to Mars and back.

Two European stations undertaking space-relevant research are based in the Antarctic peninsula. The Italian/French Concordia base which supports remote isolation studies and is open for research groups from all over Europe. ESA's Directorate of Human Spaceflight uses Concordia's special environment to prepare for future human missions to the Moon or Mars, and ESA supports the French Polar Institute and the Italian Antarctic Programme in medical monitoring, operational validation of life-support technologies and psychological training. The Rothera Research Station is the principal British Antarctic Survey logistics centre for support of Antarctic field science.

In addition to the facilities above **COMEX's Hyperbaric Experimental Centre** in France features hypobaric and hyperbaric chambers which can be used for the qualification of equipment, and human intervention methods in hostile environments; a Neutral Buoyancy Facility (Pool) for undertaking Microgravity, Lunar or Martian gravity training; and a cellular biology laboratory. **TNO, based in Soesterberg, the Netherlands** have different **Climate Chambers** for undertaking thermal physiology research of humans in extreme environments.

2.4.3 Centrifuges

The use of hypergravity devices plays an important role in many research areas from physiology and biology to materials science and technology [23, 25, 65, 70,

73, 74, 79]. Using human centrifuges in astronautics play an important role in astronaut selection by testing the capability to withstand hypergravity as a simulation to conditions of launch and re-entry. During the first minutes of a Shuttle launch for example, astronauts are exposed to accelerations of 4 g max. in a supine position. Centrifuges further hold applicability to human physiology research for example with testing centrifugation as a means of countering bone mass loss. This type of research can also impact on vestibular-related disorders in space and on Earth. Taking non-human centrifuges into account these devices are a key element in, for example, plant and cellular biology for determining the gravisensing mechanisms and thresholds in plants which could impact on cultivation processes on Earth and for future human spaceflight exploration missions beyond low-Earth orbit.

2.4.3.1 Human Centrifuges

A number of human centrifuges exist around Europe for undertaking human physiology research in hypergravity. DLR, in Cologne, Germany and the Medes Space Clinic, in Toulouse, France have similar short-arm human centrifuges (Fig. 2.19) which go up to 6 g and are used in countermeasure studies for astronauts. The centrifuges can accommodate two reclining and two seated subjects. Additional human centrifuges are in place at the Netherlands Aeromedical Institute, in Soesterberg, the Netherlands which can hold up to 175 kg in payload mass, and at the Karolinska Institute, in Stockholm, Sweden which can expose up to 300 kg of payload mass to up to nine times gravity (15 times gravity for just equipment). TNO in Soesterberg, the Netherlands also has a facility called Desdemona [8]. This is a

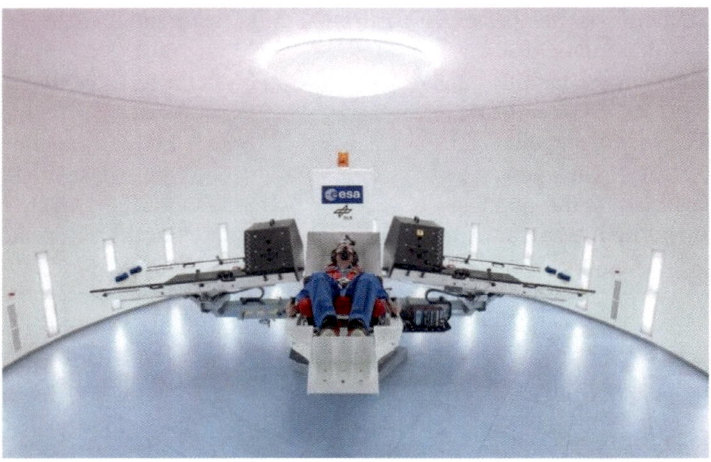

Fig. 2.19 The short-arm human centrifuge at DLR in Cologne (Source: DLR/Markus Steur)

Fig. 2.20 The ESA Large Diameter Centrifuge (LDC) at ESA-ESTEC, Noordwijk, the Netherlands (Source: ESA)

six degrees-of-freedom motion base that is capable of rotation in three axes, linear motion along an 8 m track, and sustained centrifugation up to 3 g.

2.4.3.2 Non-human Centrifuges

Non-human centrifuges around Europe vary considerably depending on their purpose. Larger radius centrifuges are in situ at the Academic Medical Center, in Amsterdam, the Netherlands and at ESA/ESTEC, Noordwijk, the Netherlands providing 8 g and up to 20 g environments respectively for a variety of biological, biotechnological, biochemical, physical, material and fluid science, geology and plasma physics experiments. The ESTEC Large Diameter Centrifuge (LDC) [80] can hold experiments (up to 80 kg) lasting from 1 min to 6 months (Fig. 2.20). DLR also has the smaller NIZEMI Centrifuge which consists of a Slow Rotating Centrifuge Microscope for hyper-g experimentation up to 5 g and allowing for observation of small organisms, and the Dutch Experiment Support Centre has the Medium sized Centrifuge for Acceleration Research (MidiCAR) which is a dedicated cell/tissue culture centrifuge in which samples may be exposed to accelerations up to 100 g.

2.4.4 Human-Rated Linear and Angular Accelerators

In a similar way to the centrifuges listed above, the Human-rated Linear and Angular Accelerators play an important role in understanding the activity and

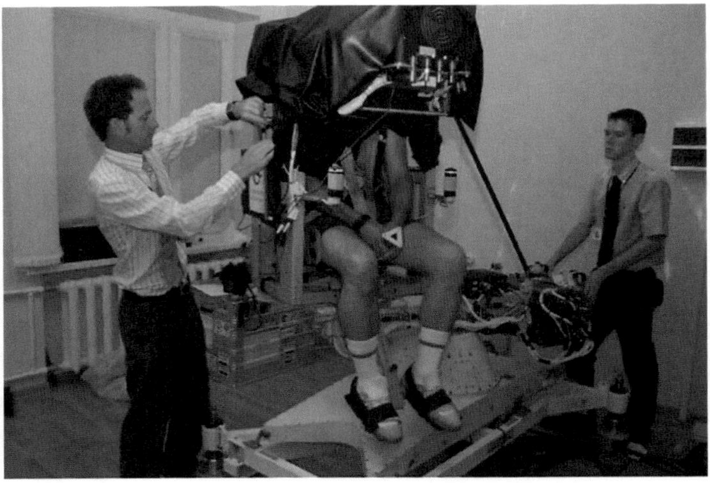

Fig. 2.21 Roscosmos cosmonaut Yuri Malenchenko seated on the visual and vestibular investigation system in star city in Moscow, undergoing a final check before centrifugation (Source: GCTC Moscow)

mechanisms underlying the vestibular system in weightlessness [6, 9, 10, 24, 48, 66, 67]. With the vestibular system relying on gravitational stimuli on Earth in order to hold proper posture and balance, the understanding of the mechanisms that underly the lack of these stimuli in space can help to provide an insight into the problems associated with special awareness and orientation in space as well as balance disorders on Earth.

TNO, in Soesterberg, the Netherlands has a vestibular laboratory with a 3-D rotating chair, Linear Track (the ESA Space Sled), Tilting Room, and Ship Motion Simulator while the **Medes Space Clinic, Toulouse, France has a Visual and Vestibular Investigation System** (Fig. 2.21)**,** ESA-developed for NEUROLAB for investigating the role of the inner ear in detecting changes in motion and orientation. Outside of these two facilities the **Centre for Human Sciences Impact Facility in Farnbourough, UK,** has a **Deceleration Track** for the Impact testing.

2.4.5 Clinostats, Free Fall Machines, and Random Positioning Devices

These devices provide/simulate weightlessness in a variety of ways [11, 13, 52–54, 68, 75, 78] for different periods of time. A clinostat (Fig. 2.22 left) uses rotation to negate the effects of gravitational pull on plant growth (gravitropism) and development (gravimorphism). It has been used to study the effects of simulated

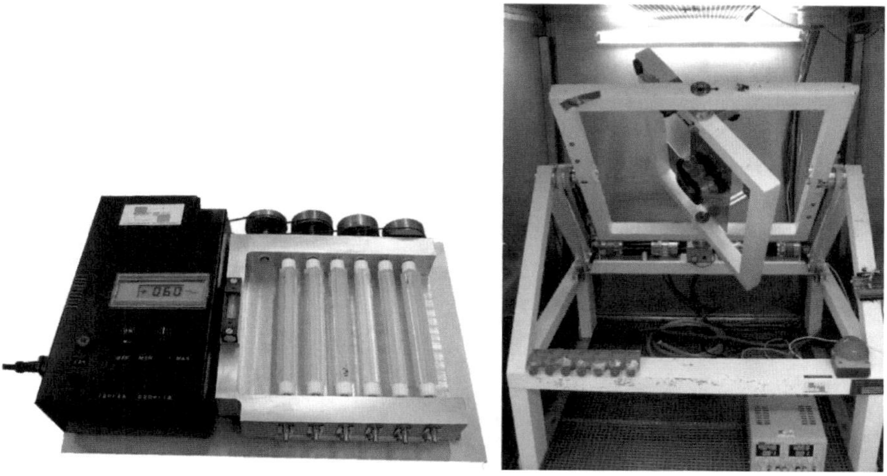

Fig. 2.22 *Left* a classical fast rotating clinostat. *Right*: the random positioning machine developed by Dutch Space (Source: DESC, J. van Loon)

microgravity on e.g. cell cultures and animal embryos. The core of the Free Fall Machine is a vertical bar which guides the experiment while it goes through its free fall cycles. After each cycle the experiment experiences a far shorter acceleration cycle at multiple g levels to return the experiment to its initial position. Random Positioning Machines (Fig. 2.22 right) rotate in a way to simulate weightlessness by removing the effect of gravity in any specific direction. Clinostat facilities are available at DLR in Cologne, the DESC Laboratory at the Free University of Amsterdam, and at the Space Biology Group of the Swiss Federal Institute of Technology in Zurich. The two latter institutions have the availability of Free Fall Machines and Random Positioning Machines.

2.4.6 Telemedicine, Behaviour and Metrics

Telemedicine can be defined as the delivery of healthcare services, where distance is a critical factor, by healthcare professionals using information and communications technologies for the exchange of valid information for diagnosis, treatment and prevention of diseases and injuries, research and evaluation, and for the continuing education of healthcare providers, all in the interest of advancing the health and their communities (WHO, 1997). The rapid advances of Information and Communications Technology offers the possibility of improving health services and making the best use of limited and valuable resources.

Within the areas of telemedicine, behaviour and metrics capabilities include: the **Telemedicine Portable Workstation developed at MEDES, Toulouse, France**, which collects biomedical patient data and can transmit them to a medical expert for a first aid medical consultation or a second opinion advice; **3-D body scanning, at TNO, Soesterberg, the Netherlands** for digitally recording exact shape and body dimensions of humans and objects which can be used in computer-aided design to construct made-to-measure apparel for example; and **Usability Engineering, at TNO Soesterberg** where equipment and human-factors know-how are available for the design and test of user interface for space applications. The services include the specification of user requirements, interface prototypes (*e.g.* storyboards), expert reviews and user tests (in the lab, on location or remotely).

2.4.7 Integrated Bio-Processing, Tissue Engineering

Developments in the multidisciplinary field of tissue engineering have yielded a novel set of tissue replacement parts and implementation strategies. Scientific advances in biomaterials, stem cells, growth and differentiation factors, and biomimetic environments have created unique opportunities to fabricate tissues in the laboratory from combinations of engineered extracellular matrices ("scaffolds"), cells, and biologically active molecules. Among the major challenges facing tissue engineering is the need for more complex functionality, as well as functional and biomechanical stability in laboratory-grown tissues destined for transplantation.

The Charité Institute for Transplantation and Organ Replacement, which forms part of the Charité, Medical Faculty of the Humboldt University in Berlin offers a scientific environment in the fields of tissue engineering, biomedical technology and transplantation medicine.

Equipment and know-how are available at **Bio-up, at the Blaise Pascal University, in Clermont-Ferrand, France** for the experimental determination of capabilities of various biological transformations, including aerobic and photosynthetic cultures of microorganisms, which are of interest in Life Support Systems for long duration space missions.

The **ERISTO ("European Research in Space and Terrestrial Osteoporosis")** project is funded by ESA, national space agencies and the ERISTO partners within the frame of the ESA Microgravity Application Promotion programme. The objectives of ERISTO are to develop innovative models of osteoporosis either using the space environment to provide "mechanical stress free" experimental conditions and to improve diagnosis, prevention and treatments of this disease. The ERISTO team has the expertise and provides access to facilities and services covering the main field of research in bone remodelling and osteoporosis [2, 12, 56, 57, 59, 62].

The main services and the expertise provided by ERISTO are: Innovative analytical tools, in particular, a system able to measure bone micro-architecture and calculate bone strength *in-vivo*. ERISTO also masters the main tools required to cultivate and study bone cells and tissues in a controlled environment; provide *in vitro* and *in-vivo* models (for example cultivation of *ex-vivo* bone cells) and provide access to ERISTO partner facilities and to new facilities developed within the project.

2.4.8 *Magnetic Resonance Facilities*

Magnetic resonance techniques are extremely useful within spaceflight-related programmes and projects for determining/imaging some of the physiological effects on soft and hard body tissues. This can either be associated with actual spaceflight missions or in simulated weightlessness such as in bed rest studies. For magnetic resonance techniques related to spaceflight there are facilities within the University of Trieste within the Dept. of Biochemistry, Biophysics and Macromolecular Chemistry (at the Cattinara Hospital in Trieste) and at the Muscle Lab of DLR's Physiology Laboratory. The Muscle Lab is available to in-house researchers and external scientists (under certain circumstances).

2.4.9 *Movement Analysis, Physical and Skills Training*

2.4.9.1 CAR (Centre d'alt Rendiment) Barcelona, Spain

The CAR center has been specifically designed to support the improvement of performances of top athletes and to characterize the physiological and general training conditions contributing to such improvement. The Olympic Training Centre is providing all necessary equipment and know-how that is necessary for the improvement of performances of top athletes. In addition to sports, educational and residential facilities, CAR offers services in biomechanics including 2D and 3D videographic analysis of movement and training, and strength development control through electrical activity (electromyography); services in physiology such as analysis of body composition, lactic acid and pH level tests, strength/force testing (dynamometry), MRI assessment of body/weight distribution and Muscular metabolism study; as well as psychology, nutrition, and physical training and evaluation. CAR is a public company of the Government of Generalitat de Catalunya with an agreement with Spanish Sports Council.

2.4.10 *Additional Animal Physiology Facilities*

In addition to the ERISTO facilities mentioned previously (encompassing mice and rat research), the developmental space biology group at the University of Nancy offers help for preparation and conditioning of embryos, larvae or adults for a space flight. In their laboratory, the reproduction and rearing of model amphibian

Pleurodeles waltl [1, 26, 29, 72] can be routinely performed. In Germany the AquaHab, owned by OHB-System AG, in Bremen is an aquatic research module based on hardware developed for DLR and flown successfully in space. Dedicated mainly to ground based research, Aquahab is supported by complex technology and a laboratory facility for operating the modules including a standard biochemical laboratory environment, hardware development and test laboratory, as well as in-house hatchery capabilities and expert personnel support.

2.4.11 Additional Plant Physiology Facilities

In addition to the facilities and equipment mentioned previously (Centrifuges, Magnetic Resonance, Clinostats, Aquahab etc.) principal facilities for plant physiology research [19–21, 77] related to space applications are:

2.4.11.1 Plant Biocentre and Norwegian User Support and Operations Centre, University of Science and Technology, Trondheim, Norway

The Plant Biocentre has the necessary laboratory facilities for cultivation and analysis for plant biology research, while the User Support and Operations Centre for the European Modular Cultivation System (discussed previously), is used for testing the design concept of planned experiment hardware and selected plant material before the performance of experiment in space.

2.4.11.2 Multispectral Plant Imaging, University of Ghent, Belgium

At this facility which is part of the Department of Molecular Genetics the automated thermography setup permits time-lapse thermal imaging of leaves of multiple plants. Combined thermographic and video imaging is available to obtain spatial correlation of visible and thermal stress symptoms as a function of time. On the software side, automated generation of overview images and movies for rapid visualisation of changes was developed.

2.4.12 Magnetic Levitation

Every element (in the periodic system) is magnetic from the smallest effect know as diamagnetism which is a property of elements like copper and carbon to the largest magnetic effect known as ferromagnetism, which in our daily lives is the most common form of magnetism and found in elements such as iron, cobalt, and nickel. Ferromagnetism is about 109 times larger compared to diamagnetism. Diamagnetic materials, need a much higher gradient magnetic field strength in order to be levitated (and simulate weightlessness). This can be carried out on organic materials or even living biological samples [5, 7, 81]. Such facilities are available at the **High Field Magnet Laboratory of the University of Nijmegen, in the Netherlands** with proposals processed by the Dutch Experiment Support Centre, or via an international 'Announcements of Opportunities' generally issued annually by ESA or other space agencies. You may also apply via an unsolicited proposal to ESA via the 'fast track' Continuously Open Research Announcements or via the EC FP-7 program EuroMagNET II: Research Infrastructures for High Magnetic Field in Europe. Similar facilities for fluid sciences, material sciences, and fundamental physics research are present at the French government funded **Alternative Energies and Atomic Energy Commission, in Grenoble, France**.

2.4.13 Biotechnology and Life Support Systems

With the prospect of future human spaceflight exploration missions, the need to reduce launch/upload mass is an essential part in mission planning. Biotechnology could have a significant effect on this with the development of regenerative life support systems, thus reducing the need to launch, for example, large quantities of drinking water or atmospheric gases. The **Micro-Ecological Life Support System Alternative (MeliSSA),** which is a collaborative project managed by ESA is conceived as a micro-organisms and higher plants based ecosystem intended as a tool to gain understanding of the behaviour of artificial ecosystems (see also Sect. 3.2.9 of this book). Based on the principle of an aquatic ecosystem the second-generation **MeLISSA pilot plant at the University Autònoma of Barcelona, Spain** is testing regenerative life support system technologies. In addition the **Sub-department of Environmental Technology of the University of Wageningen in the Netherlands** has facilities to study (microbiologically mediated) conversions of organic and inorganic matter under simulated planetary atmosphere environments. The expertise of the facility can also support research on the development of more efficient and environment friendly technologies needed for space exploration.

2.4.14 Extreme Environments

Simulation of environmental conditions in space [60, 61], and on other planets and orbital bodies can be an important precursor in astrobiological experiments in order to determine the survivability of different species under these conditions. It can also help in the planning of future missions, by testing different equipment and technologies to verify that it can deal with the conditions in which it has to function. **The Planetary (Mars) simulation facilities at the DLR "Mars-complex" in**

Cologne provide the following simulated Martian environment for physical and exobiological studies: atmosphere, UV-radiation climate, surface/subsurface temperature, controlled with regard to diurnal/seasonal fluctuation. Announcements of Opportunity are made at a European level and about 50% of the facility resources can be made available to visiting scientists. At **the Organics under Simulated Interstellar Conditions (OSIC) at the University of Leiden in the Netherlands** there is a model chamber dedicated to the study of carbonaceous material in ultra-high vacuum at low temperature and under UV irradiation.

2.4.15 Radiation Testing

Similar to the section on extreme environments above, facilities testing the effects of radiation in different environments are a vital part of planning for future spaceflight missions, with the testing of different components, equipment and technologies verifying that it can deal with exposure to the different levels of radiation in orbit in which it has to function. Developments in this area can also have an impact in areas of medicine such as within heavy ion therapies for cancer treatment.

Numerous facilities are present around Europe: ESA operates **Internal Radiation Test Facilities at ESTEC, Noordwijk, the Netherlands,** which has a Co-60 gamma source for testing as well as the CASE (Californium-252 Assessment of Single-event Effects) laboratory test system, which is an alternative to the conventional heavy ion accelerator; The **GSI Helmholtz Centre for Heavy Ion Research, Darmstadt, Germany** can simulate cosmic radiation, in particular the galactic cosmic radiation, and has the accelerators UNILAC and SIS-18 that deliver ion beams of high quality of many chemical elements (including iron) in the energy range of 1 MeV–1 GeV per nucleon and can accelerate light ions (up to neon) to 2 GeV per nucleon; **DLR's Integrated Space Environment Factors Simulator – KOBE in Berlin Germany** offers interstellar and planetary environment simulation such as high vacuum, Martian climate conditions, solar irradiation and ultraviolet radiation; the **Grand Accélérateur National d'Ions Lourds (GANIL) in Caen, France** offers production of accelerated ions from helium to uranium of medium energy (20–100 MeV/amu) which can be used for studies in fundamental physics, but also in radiobiology and material sciences as simulating the exposition of biological systems or materials to cosmic heavy ions; the accelerator facility at the **Paul Scherrer Institute, Switzerland** is used to accelerate protons which can be used within research into the smallest fundamental constituents of matter, the investigation of innovative materials, the development of new products for medical diagnosis and unique methods of treating tumours; the **Heavy Ion Irradiation Facility, at the Centre de Recherches du Cyclotron, Louvain-la-Neuve, in Belgium** is used doe studies in Single Event Effects in collaboration with ESA; and the RADiation Effects Facility (RADEF), which is located in the Accelerator

Laboratory of the University of Jyväskylä, Finland includes beam lines dedicated to proton and heavy ion irradiation studies of semiconductor materials and devices.

2.4.16 *Fluid Science Facilities: Surface Tension*

The **SMT-laboratory, University of Genoa, Italy** is composed of a number of dedicated pieces of equipment, which allow measurement to be made from near-ambient temperatures up to 1,500°C. Measurements relevant to fluid and materials science can be made at both the liquid-gas interface and liquid-liquid interface. Special software treats, in real time, all relevant input data so that even fast interfacial phenomena, like those governed by adsorption and diffusion can be traced. Surface and interfacial tension data, both at equilibrium and in dynamic conditions are of basic importance for studies related to solidification, crystal growth, joining, detergency, foams and emulsion stability etc.

2.4.17 *Materials Science Facilities: Crystallisation*

The **Universidad Autónoma de Madrid, Spain** offers equipment for sample preparation and growth of single crystals at high temperature, in air or controlled atmosphere, melt and vapour techniques; as well as capabilities for cutting, polishing and orientation of single crystals; and methods of sample characterization techniques. The timeline for performing an experiment is 3–12 months depending on the adaptation of equipment to a required experiment.

2.4.18 *Materials Science Facilities: Solar Power*

In operation since 1991, the **Solar Furnaces in Almeria, Spain** [46, 47, 49, 63, 76] have been fully devoted to materials treatment in the framework of European Union-funded research programmes. The main components are: the mirrored, mobile heliostats, which reflect sunlight; the mirrored parabolic concentrator onto which sunlight from the heliostats is reflected; a louvered shutter to control sunlight concentration; and a movable test table, which in turn concentrates sunlight onto the focal spot where specimens are held. **DLR has a 25 kW Solar Furnace, a high power radiation source (20 kW) and further solar test facilities in Cologne, Germany.** Well equipped laboratories, workshops and simulation tools in Stuttgart and Cologne allow for thermal, chemical, optical R&D activities as well as system analyses.

2.4.19 Materials Science Facilities: Wind Tunnels

At **CORIA, at the University of Rouen, France** three wind tunnels have been built to **simulate re-entry conditions** of different planetary atmospheres. Numerous measurement techniques, in particular optical diagnostics, have been developed to study high enthalpy flows and supersonic plasma flow can also be generated. A similar wind tunnel exists at the **Von Karman Institute Sint-Genesius-Rode, Belgium**. Temperatures up to 10,000 K can be achieved.

At the University of Aarhus, in Denmark simulation of the Martian aerosol is performed in a unique re-circulating wind tunnel enclosed in a low pressure atmospheric chamber. Importantly such a system allows the atmosphere to be carefully controlled and monitored and the dust to be stored for long periods of time compared to flow-through systems. Typical wind speeds of 0–10 m/s can be reproduced with variable dust density. A liquid nitrogen cooling system allows the extreme low temperatures on Mars to be achieved, this also allows the low humidity to be reproduced. In the 2 years of research it has been found that dust sticks readily to any and all surfaces invariably forming aggregates as the dust sticks to itself.

References

1. Aimar, C., Bautz, A., Durand, D., et al.: Microgravity and hypergravity effects on fertilization processes in the salamander Pleurodeles waltl. Biol. Reprod. **63**, 551–558 (2000)
2. Akhouayri, O., Lafage-Proust, M.H., Rattner, A., et al.: Effects of static or dynamic mechanical stresses on osteoblast phenotype expression on three-dimensional contractile collagen gels. J. Cell. Biochem. **76**, 217–230 (1999)
3. Columbus Payloads Accommodation Handbook. Ref # COL-RIBRE-MA-O007-00. Astrium GmbH, Bremen, Germany (2001)
4. Columbus External Payloads Interface Requirements Document. ref # COL–RIBRE–SPE–0165. Astrium Gmbh, Bremen, Germany (2002)
5. Beaugnon, E., Tournier, R.: Levitation of water and organic substances in high static magnetic fields. J. Phys. III Fr. **1**, 1423–1428 (1991)
6. Bellossi, F., Clément, G., Cohen, B., et al.: EDEN: a payload dedicated to neurovestibular research for Neurolab. Acta Astronaut. **42**(1–8), 59–67 (1998)
7. Berry, M.V., Geim, A.K.: Of flying frogs and levitrons. Eur. J. Phys. **18**, 307–313 (1997)
8. Bles, W.: Desdemona: Advanced Disorientation Trainer. HSIAC publication Gateway, SD-edition Fall (2001)
9. Bles, W., van Raaij, J.L.: The tilting room and the spacelab D1 mission. Adv. ORL **42**, 13–17 (1988)
10. Bles, W., Bos, J.E., de Graaf, B., et al.: Motion sickness: only one provocative conflict? Brain Res. Bull. **47**(5), 481–549 (1998)
11. Borst, A.G., van Loon, J.J.W.A.: Technology and developments for the random positioning machine. RPM. Microgravity Sci. Technol. **21**(4), 287–292 (2009)
12. Boyde, C.A., Quarto, R., et al.: Osteoconduction in large microporous hydroxyapatite ceramic implants: evidence for a complementary integration and disintegration mechanism. Bone **24**, 579–589 (1999)
13. Briegleb, W.: Some quantitative aspects of the fast-rotating clinostat as a research tool. ASGSB Bull. **5**, 23–30 (1992)

14. Ceglia, E., Carey, W., ESA.: UIC-ESA-UM-001 European Users Guide to Low Gravity Platforms (Rev 2.0). 3. Drop Towers (2005)
15. Ceglia, E., Carey, W., ESA.: UIC-ESA-UM-001 European Users Guide to Low Gravity Platforms (Rev 2.0). 4. Parabolic Flights (2005)
16. Ceglia, E., Carey, W., ESA.: UIC-ESA-UM-001 European Users Guide to Low Gravity Platforms (Rev 2.0). 5. Sounding Rockets (2005)
17. Ceglia, E., Carey, W., ESA.: UIC-ESA-UM-001 European Users Guide to Low Gravity Platforms (Rev 2.0). 6. Foton Retrievable Capsules (2005)
18. Ceglia, E., Carey, W., ESA.: UIC-ESA-UM-001 European Users Guide to Low Gravity Platforms (Rev 2.0). 7.3 Physical Environment (2005)
19. Chaerle, L., Van der Straeten, D.: Imaging techniques and the early detection of plant stress. Trends Plant Sci. **5**, 495–501 (2000). S.C.I.: 9.350
20. Chaerle, L., Van Caeneghem, W., Messens, E., et al.: Presymptomatic visualisation of plant-virus interactions by thermography. Nat. Biotechnol. **17**, 813–816 (1999). S.C.I.: 10.117
21. Chaerle, L., De Boever, F., van Montagu, M., et al.: Thermographic visualisation of cell death phenomena in tobacco and Arabidopsis. Plant Cell Environ. **24**, 15–26 (2001). S.C.I.: 3.579
22. Chesson, R.: The ISS Operations & Exploitation Program. ESA Bulletin. **110**, 61–68 (2002)
23. Clement, G., Bukley, A.: Artificial Gravity. Springer, New York (2007)
24. Clément, G., Moore, S., Raphan, T., et al.: Perception of tilt (somatogravic illusion) in response to sustained linear acceleration during space flight. Exp. Brain Res. **138**(4), 410–418 (2001)
25. de Graaf, B., de Roo, A.J.: Effects of long duration centrifugation on head movements and a psychomotor task. J. Vestib. Res. **6**(1), 23–29 (1996)
26. Dournon, C., Durand, D.D., Tankosic, C., et al.: Further larval development, metamorphosis and reproduction of Pleurodeles waltl (urodele amphibian) born and developed up to hatching stage in microgravity. Develop. Growth & Differ **43**, 315–326 (2001)
27. Drummer, C., Gerzer, R., Baisch, F., Heer, M.: Body fluid regulation in μ- gravity differs from Earth: an overview. Pflugers Arch. **441**(Suppl), R66–R72 (2000)
28. Drummer, C., Heer, M., Joosten, M., et al.: Regulation and distribution of body fluid during a 6-day head-down tilt study in a randomized cross-over design. J. Gravit. Physiol. **7**, P-187–P-188 (2000)
29. Durand, D., Tankosic, C., Chaput, D., et al.: Feeding and rearing techniques used for larvae of Pleurodeles waltl (urodele amphibian) onboard the MIR space station. Scand. J. Lab. Anim. Sci. **27**, 193–201 (2000)
30. Ceglia, E. Ch.9: European Drawer Rack. In: Carey E.W. (ed) European Users Guide to Low Gravity Platforms. Issue 2 Rev 0. ESA Ref: UIC-ESA-UM-001. ESA, Noordwijk, The Netherlands
31. Ceglia, E. Ch.9: Biolab – Biological laboratory in Columbus. In: Carey E.W. (ed) European Users Guide to Low Gravity Platforms. Issue 2 Rev 0. ESA Ref: UIC-ESA-UM-001. ESA, Noordwijk, The Netherlands
32. Ceglia, E. Ch.9: Fluid Science Laboratory (FSL). In: Carey E.W. (ed) European Users Guide to Low Gravity Platforms. Issue 2 Rev 0. ESA Ref: UIC-ESA-UM-001. ESA, Noordwijk, The Netherlands
33. Ceglia, E. Ch.9: European Physiology Modules (EPM). In: Carey E.W. (ed) European Users Guide to Low Gravity Platforms. Issue 2 Rev 0. ESA Ref: UIC-ESA-UM-001. ESA, Noordwijk, The Netherlands
34. Ceglia, E. Ch.9: Protein Crystallisation Diagnostics Facility. In: Carey E.W. (ed) European Users Guide to Low Gravity Platforms. Issue 2 Rev 0. ESA Ref: UIC-ESA-UM-001. ESA, Noordwijk, The Netherlands
35. Ceglia, E. Ch.9: SOLAR. In: Carey E.W. (ed) European Users Guide to Low Gravity Platforms. Issue 2 Rev 0. ESA Ref: UIC-ESA-UM-001. ESA, Noordwijk, The Netherlands

36. Ceglia, E. Ch.9: Materials Science Laboratory (MSL). In: Carey E.W. (ed) European Users Guide to Low Gravity Platforms. Issue 2 Rev 0. ESA Ref: UIC-ESA-UM-001. ESA, Noordwijk, The Netherlands
37. Ceglia, E. Ch.9: Microgravity Science Glovebox (MSG). In: Carey E.W. (ed) European Users Guide to Low Gravity Platforms. Issue 2 Rev 0. ESA Ref: UIC-ESA-UM-001. ESA, Noordwijk, The Netherlands
38. Ceglia, E. Ch.9: Matroshka – Space Radiation Experiment. In: Carey E.W. (ed) European Users Guide to Low Gravity Platforms. Issue 2 Rev 0. ESA Ref: UIC-ESA-UM-001. ESA, Noordwijk, The Netherlands
39. Ceglia, E. Ch.9: Minus Eighty Laboratory Freezer for ISS. In: Carey E.W. (ed) European Users Guide to Low Gravity Platforms. Issue 2 Rev 0. ESA Ref: UIC-ESA-UM-001. ESA, Noordwijk, The Netherlands
40. Ceglia, E. Ch.9: European Technology Exposure Facility. In: Carey E.W. (ed) European Users Guide to Low Gravity Platforms. Issue 2 Rev 0. ESA Ref: UIC-ESA-UM-001. ESA, Noordwijk, The Netherlands
41. Ceglia, E. Ch.9: Atomic Clock Ensemble in Space (ACES). In: Carey E.W. (ed) European Users Guide to Low Gravity Platforms. Issue 2 Rev 0. ESA Ref: UIC-ESA-UM-001. ESA, Noordwijk, The Netherlands
42. Ceglia, E. Ch.9: Pulmonary Function System (PFS). In: Carey E.W. (ed) European Users Guide to Low Gravity Platforms. Issue 2 Rev 0. ESA Ref: UIC-ESA-UM-001. ESA, Noordwijk, The Netherlands
43. Ceglia, E. Ch.9: Flywheel Exercise Device (FWED). In: Carey E.W. (ed) European Users Guide to Low Gravity Platforms. Issue 2 Rev 0. ESA Ref: UIC-ESA-UM-001. ESA, Noordwijk, The Netherlands
44. Ceglia, E. Ch.9: European Modular Cultivation System. In: Carey E.W. (ed) European Users Guide to Low Gravity Platforms. Issue 2 Rev 0. ESA Ref: UIC-ESA-UM-001. ESA, Noordwijk, The Netherlands
45. Ceglia, E. Ch.9: MARES. In: Carey E.W. (ed) European Users Guide to Low Gravity Platforms. Issue 2 Rev 0. ESA Ref: UIC-ESA-UM-001. ESA, Noordwijk, The Netherlands
46. Farinha Mendes, J., Collares-Pereira, M., Martinez, D., et al.: Testing results of a second stage concentrator designed for the solar furnace of plataforma solar dAlmería. J. Phys. IV **9**, Pr3-569–Pr3-574 (1999)
47. Fernandez, BJ., Lopez, V., Vázquez, AJ., et al.: Cladding of Ni superalloy powders on AISI 4140 steel with concentrated solar energy. Solar Energy Materials and Solar Cells. **53**, 153–161 (1998)
48. Groen, E.L., Bos, J.E., de Graaf, J.E., et al.: Otolith signal processing and motion sickness. Ann. NY Acad. Sci. **871**, 406–409 (1999)
49. Guerra Rosa, L., Cruz Fernandes, J., Amaral, P.M., et al.: Photochemically promoted formation of higher carbide of molybdenum through reaction between metallic molybdenum powders and graphite powders in a solar furnace. Int. J. Refract. Met. Hard Mater. **17**, 351–356 (1999)
50. Heer, M.: Influence of dietary increased NaCl intake on body fluid and electrolyte metabolism. Thesis/Dissertation, University of Bonn, Institute of Nutrition Science, Bonn, Germany (1996)
51. Heer, M., Drummer, C., Baisch, F., et al.: Effects of head-down tilt and saline loading on body weight, fluid, and electrolyte homeostasis in man. Acta Physiol. Scand. Suppl. **604**, 13–22 (1992)
52. Hemmersbach-Krause, R., Briegleb, W., Häder, D.P., et al.: Orientation of Paramecium under the conditions of weightlessness. J. Euk. Microbiol. **40**, 439–446 (1993)
53. Hensel, W., Sievers, A.: Effects of prolonged omnilateral gravistimulation on the ultrastructure of statocytes and on the graviresponse of roots. Planta **150**, 338–346 (1980)
54. Hoson, T., Kamisaka, S., Masuda, Y., et al.: Changes in plant growth process under microgravity conditions simulated by a three-dimensional clinostat. Bot. Mag. **105**, 53–70 (1992)

55. ISS Intergovernmental Agreement: Agreement Among The Government Of Canada, Governments Of Member States Of The European Space Agency, The Government Of Japan, The Government Of The Russian Federation and The Government Of The United States Of America Concerning Cooperation On The Civil International Space Station (1998)
56. Kohlbrenner, H.S., Laib, A., Rüegsegger, P.: A 3D microtomographic system with stacked fan-beam geometry. Nucl. Instrum. Methods Phys. Res. **A443**, 531–539 (2000)
57. Kon, E., Muraglia, A., Corsi, A., et al.: Autologous bone marrow stromal cells loaded onto porous hydroxyapatite ceramic accelerate bone repair in critical size defects of sheep long bones. J. Biomed. Mater. Res. **49**, 328–337 (2000)
58. Krige, J., Russo, A.: A History of the European Space Agency 1958–1987, vol. I + II. R.A. Harris (ed.), ESA SP-1235, ESA Publications Division, Noordwijk, The Netherlands (2000)
59. Laib, A., Barou, O., Vico, L., Lafarge-Proust, M.H., et al.: 3D micro-computed tomograhy of trabecular and cortical bone architecture with application to a rat model of immobilization osteoporosis. Med. Biol. Eng. Comput. **38**(3), 326–332 (2000)
60. Lura, F., Troppenz, E., Lötzke, HG.: Creation, calculation and testing of the thermal control system for the CASSINI cosmic dust analyser. In: Proceedings of the 26th International Conference on Environmental Systems, Monterey (1996, July)
61. Lura, F., Hagelschuer, D., Abraimov, V.V.: KOBE – The simultaneous simulation of space environment effects for the investigation of spaceborne material properties. In: Proceedings of the 3rd International Symposium on Environmental Testing for Space Programmes. ESA - ESTEC, Noordwijk (1997, June)
62. Marcacci, M., Kon, E., Zaffagnini, S., et al.: Reconstruction of extensive long bone defects in sheep using porous hydroxyapatite sponges. Calcif. Tissue Int. **64**, 83–90 (1999)
63. Martınez, D., Rodriguez, J.: Surface treatment by concentrated solar energy: The solar furnace at the "Plataforma Solar de Almerıa". In: Surface modification technologies XI proceedings of the Tenth International Conference on Surface Modification Technologies, Paris, France, September 8–10, 1997. Editors (proceedings) T.S. Sudarshan, M. Jeanding, K.A. Khor. Published 1998 by Institute of Materials in London, UK. 441–447 (1998)
64. Minster, O., Innocenti, L., Mesland, D.: Looking at Science on Board Eureca, ESA BR – 80, pp. 3–6. ESA Publications Division, ESTEC, Noordwijk (1993)
65. Mitchell, R.J.: The eleventh annual R.M. Hardy Keynote address, 1997: centrifugation in geoenvironmental practice and education. Can. Geotech. J. **35**, 630–640 (1997)
66. Moore, S.T., Clément, G., Raphan, T., et al.: The human response to artificial gravity in a weightlessness environment: Results from the Neurolab centrifugation experiments. In: El-Genk, M.S. (ed.) Space Technology and Applications International Forum 2000, pp. 206–211. American Institute of Physics, College Park, MD (2000)
67. Moore, S., Clément, G., Raphan, T., et al.: Ocular counterrolling reflex (OCR) induced by centrifugation during orbital space flight. Exp. Brain Res. **137**, 323–335 (2001)
68. Muller, H.J.: Approximation to a gravity-free situation for the human organism achievable at moderate expense. Science **128**, 772 (1959)
69. NASA Johnson Space Center: International Space Station Familiarization. TD9702A (1998)
70. Nooij, S.: Vestibular adaptation to an altered gravitational environment, Consequences for spatial orientation. 978-90-9022982-9 (2008)
71. Norsk, P., Christensen, N.J., Bie, P., et al.: Unexpected renal responses in space [letter] [In Process Citation]. Lancet **356**, 1577–1578 (2000)
72. Oukda, M., Francois, M., Membre, H., et al.: Crystallographic and chemical composition of otoconia in the salamander Pleurodeles waltl. Hear. Res. **132**, 85–93 (1999). Rank 2/29, Otorhinolaryngology, IF 1,598 – CHL 6,9
73. Regel, L.L., Wilcox, W.R. (eds.): Materials Processing in High Gravity. Plenum Press, New York (1994)
74. Rohdin, M., Linnarsson, D.: Differential changes of lung diffusing capacity and tissue volume in hypergravity. J. Appl. Physiol. **93**, 935 (2002)

75. Schwarzenberg, M., Pippia, P., Meloni, M.A., et al.: Signal transduction in T lymphocytes – a comparison of the data from space, the free fall machine and the random positioning machine. Adv. Space Res. **24**, 793–800 (1999)
76. Shohoji, N., Guerra Rosa, L., Cruz Fernandes, J., et al.: Catalytic acceleration of graphitisation of amorphous carbon during synthesis of tungsten carbide from tungsten and excess amorphous carbon in a solar furnace. Mater. Chem. Phys. **58**, 172–176 (1999)
77. Van Der Straeten, D., Chaerle, L., Sharkov, G., et al.: Salicylic acid enhances the activity of the alternative pathway of respiration in tobacco leaves and induces thermogenicity. Planta **196**, 412–419 (1995). S.C.I.: 2.977
78. van Loon, J.J.W.A.: Some history and use of the random positioning machine, RPM, in gravity related research. Adv. Space Res. **39**, 1161–1165 (2007)
79. van Loon, J.J.W.A., Tanck, E., van Nieuwenhoven, F., et al.: A brief overview of animal hypergravity studies. J. Grav. Physiol. **12**(1), 5–10 (2005)
80. van Loon, JJWA., Krause, J., Cunha, H., et al: The large diameter centrifuge, LDC, for life and physical sciences and technology. In: Proceedings of the 'Life in Space for Life on Earth Symposium', Angers, 22–27 June 2008. ESA SP-663 (2008)
81. Wunenburger, R., Chatain, D., Garrabos, Y., et al.: Magnetic compensation of gravity forces in (p−) hydrogen near its critical point; application to weightlessness conditions. Phys. Rev. E **62**, 469–476 (2000)

Chapter 3
Areas of Research

Jessica Aceto, Joan Albiol, Audrey Berthier, Daniel Beysens, Johannes Boonstra, Silvia Bradamante, Christian Chauveau, Alexander Chouker, Gilles Clément, Alain C. Colige, Bernard Comet, Volker Damann, Marco Durante, Pascale Ehrenfreund, Hans Fecht, Rüdiger Hampp, Ruth Hemmersbach, Bob Hockey, Eberhard Horn, Charles A. Lambert, Natalie Leys, Cesare Lobascio, Jack J.W.A. van Loon, Stefano Mancuso, Francisco Javier Medina, Greg Morfill, Marc Muller, Betty V. Nusgens, Maria Antonietta Perino, Günther Reitz, Hanna Rothkaehl, Gro Mjeldheim Sandal, Laurence Vico, and Frances Westall

Abstract This chapter introduces the main topics of research that have benefited so far from the space environment (reduced gravity, ambient radiation, vacuum, etc.), and provides an outlook for future research development. By convention, it is split into two fields: physical sciences/engineering and life sciences.

Physical science and engineering studies can be further divided into subfields; however, they quite often overlap (e.g. fluids). They range from very fundamental studies such as tests of special and general relativity, to the variety of fluid disciplines such as combustion and materials sciences, which are more application related. The space environment itself is also investigated.

Life sciences address the impact of microgravity and radiation on single cells, plants, animals and finally humans. There is a discussion of how life is affected by the space environment, and how we can make use of this environment to learn about some basic processes in life and how it might have developed on Earth. Leaving Earth for long-duration missions to Mars, for example, requires a sound understanding of how to deal with such a very demanding mission both with respect to crew survival and operations as well as to the technology to support human life on such missions.

D. Beysens et al. (eds.), *Laboratory Science with Space Data*,
DOI 10.1007/978-3-642-21144-7_3,

3.1 Physical Sciences

3.1.1 Fundamental Physics

Greg Morfill

Max-Planck-Institut für extraterrestrische Physik, Giessenbachstrasse,
85748 Garching, Germany

"Fundamental Physics" is somewhat of a misnomer for this chapter, since a great deal of fundamental physics research is also conducted in the research topics covered by the other chapters in this section. Also, there is a certain amount of overlapping interest – for instance in the field of "complex plasma physics" various strong coupling phenomena ranging from phase transition, phase separation, electrorheology to surfaces and self-organisation phenomena are being studied – albeit at the discrete "atomistic" level. In fluid physics and combustion we find the same topics occurring – with research conducted here at the "continuum" level. There are many other examples which serve to illustrate the fact that various research topics are complementary – a fact that has not yet been sufficiently explored and where it is hoped that this book will enhance such cross-fertilisation amongst the fields in future. With these preliminary remarks in mind let us proceed with the individual topics:

- Quantum communication for space experiments: Space QUEST
- Quantum gases/BEC in space
- Space optical clocks
- Atom interferometry sensors for space applications
- Critical point studies in complex plasmas
- Solidification of colloids in Space: Structure and dynamics of crystal, gel and glassy phases

In addition, some topics range from Lorentz invariance, the equivalence principle, soft matter (colloid, physics, granular media, cosmic dust physics, complex plasmas) to diamond formation. Such a broad approach – promising giant steps in our understanding of physics – was last seen at the beginning of the twentieth century. Today the "enabling factor" is the availability of research under microgravity conditions.

3.1.1.1 Fundamental Interactions – Quantum Physics in Space-Time

It is a really unexpected development (and nobody could have predicted this 10–15 years ago) that so many issues in fundamental physics require research in space – to acquire the higher precision, resolution etc. needed for the next breakthrough in our understanding. This covers, and here is another surprise, every timescale from the largest to the smallest, i.e., from relativity to mesoscopic phenomena, to the quantum world [21, 22, 68, 226, 239]:

One of the major puzzles – perhaps even **the** major puzzle in physics – is the incompatibility between "**General Relativity Theory**" and "**Quantum Theory".** Both theories have been tested and verified to typically one part in 10^{10} quantitatively – and must be regarded as very sound. Nevertheless, they are incompatible. A great deal of research effort is spent in trying to understand this, but so far no convincing explanation has been forthcoming. One possible resolution of this puzzle is self-gravity, which could destroy the particle wave function. Experiments to test this (e.g. massive particle interferometry) need microgravity.

Another major question concerns the fundamental constants – e.g. the gravitational constant, the fine structure constant, Planck's constant, the elementary charge, proton/electron mass, speed of light etc. Are these "constants" really universally constant, or do they vary with time when the timescale is the age of the universe? Any possibility of testing this requires enormously precise and stable clocks. These are usually based on atomic or optical processes. Comparisons can provide new thresholds of constancy or perhaps even measure possible time effects. Stable clocks again require microgravity.

Then there is the issue of "gravitational mass" and "inertial mass" – as discussed in the famous "equivalence principle". Are they really the same? And how precisely can we measure the predicted gravitational redshift? Such experiments can only be conducted in space if we wish to push the limits of detection to new records.

And last, but not least, there is the topic of "mesoscopic quantum states". This concerns Bose-Einstein condensates (BEC) of comparatively huge (billions of elementary) masses, the interactions between such mesoscopic quantum states and the possible effect of self-gravitation and quantum entanglement. Such massive BECs require microgravity in order to grow (and cool) them. On Earth they cannot be trapped for long enough.

Whilst the issues mentioned above are at the core of contemporary physics, and whilst they may appear very academic, let us not forget that, for instance, general relativity is essential to ensure proper working of all navigation systems (we can buy this very application now even in supermarkets) and that quantum theory is at the heart of the multi-billion dollar semiconductor industry, without which we would not be able to enjoy our present level of communication, medical diagnostics, transportation etc. Although fundamental research may take some time to produce tangible benefits there is no doubt that such fundamental research is worthwhile, or to put it more succinctly – absolutely essential.

Many ingenious experiments have already been devised (and carried out) to address the topics mentioned above and to push our frontier of understanding further. It seems clear, however, that the next level of quantitative advances will require microgravity, i.e. research in space, and for that purpose a number of projects are at various stages of development and preparation.

The following is a summary of different statements from the literature.

3.1.1.2 Motivations for Testing General Relativity

General Relativity rests on the Einstein Equivalence Principle which consists of three parts:

1. Universality of Free Fall, also called the Weak Equivalence Principle
2. Local Lorentz Invariance, which implies the local validity of Special Relativity
3. Local Position Invariance, which implies the universality of the gravitational red shift

Two major reasons for testing the theory (apart from the intrinsic value of such tests) are:

- Theories attempting the unification of all forces predict a violation of the Equivalence Principle.
- String theory predicts an additional scalar gravitational field, which leads to time- and position-variation of the fundamental constants, such as the fine structure constant

Furthermore, these theories also predict a deviation from the Newtonian gravitational 1/r potential.

3.1.1.3 Experimental Tests of Special and General Relativity in Space

One of the foundations of Special Relativity (SR) is the Local Lorentz Invariance (LLI). According to this principle, the outcome of any local test experiment is independent of the velocity of the freely falling apparatus. One test of this postulate to an unprecedented accuracy level would be the utilization of different frequency standards. These include clocks based on ultra-stable high-finesse resonators, atomic clocks in the optical domain while performing both on-board and space-to-ground comparisons as a function of the orbital motion of the spacecraft. Depending on the particular scheme, the set-up will be sensitive to violations of Einstein's theory of Special Relativity induced by variations of the speed of light with the orientation or the velocity of the local reference frame (tests of the isotropy and constancy of the speed of light). Measurements would allow the estimation of key parameters in alternative theories.

As a direct consequence of Einstein's Equivalence Principle, identical clocks placed at different positions in stationary gravitational fields will experience a frequency shift that, in the frame of post-Newtonian approximation (PPN) is proportional to the difference of the Newtonian potential at the positions of the clocks. Space-to-ground and on-board comparisons of atomic clocks in the optical domain can be used to accurately test Einstein's prediction, to verify the universality of the gravitational redshift and, at the same time, to detect possible time variations of fundamental constants. Such experiments would provide a comprehensive test of the Local Position Invariance (LPI) principle, which together with the Weak Equivalence Principle and Local Lorentz Invariance, constitutes the basis of Einstein's theory of General Relativity.

The following experiments could be performed:

1. Test of the isotropy of light propagation
2. Test of the independence of the speed of light from the velocity of the laboratory reference frame
3. Measurement of the speed of light by space-to-ground time and frequency transfer
4. Measurement of the relativistic time-dilation effect
5. Absolute measurement of the gravitational redshift
6. Test of the universality of the gravitational redshift
7. Search for time variations of the fine structure constant (Fig. 3.1)

3.1.1.4 Soft Matter Physics

"Soft matter" – a name given by Nobel Prize Laureate Pierre-Gilles de Gennes to a class of substances (e.g. polymers, colloids, gels, foams etc.) that exhibit macroscopic softness and whose structure and dynamics is not governed by quantum effects (e.g. mesoscopic and supra-molecular materials and material assemblies) constitute an important industrial area which is still growing in scope daily. We will specifically focus in this chapter on two "recent additions" to this field of soft matter – "complex plasmas" and "granular matter". In Sect. 3.1.2, foams and emulsions as well as complex fluids will be discussed in detail.

Here we merely wish to point out that soft matter is a broad field covering physics, chemistry and biology – with applications as disparate as paints, new and extreme materials, functionalized (bio) surfaces etc.

The "space connection" again stems from the simpler gravity-free environmental conditions. Under low gravity some systems are easier to produce, fragile structures can survive, processes such as convection are absent and therefore cannot inhibit delicate structure formation, self-organisation and dynamic processes. These comments should not be seen as advocating "production in space", rather the aim is – as in other cases mentioned later – to obtain a better understanding of fundamental processes and then to transfer this new-found knowledge to processes on Earth.

3.1.1.5 Complex Plasma Physics

Experiments in complex plasma physics have been conducted on the ISS since the very beginning, a period covering 10 years so far. During this long period the research focus has evolved considerably [76, 182, 248, 259].

In the early years the emphasis was on researching the properties of this "new state of matter" – the structure of plasma crystals, propagation of waves, domain boundaries, dislocations, crystallization fronts, melting etc. – all at the "atomistic" level of the motion of individually resolved interacting micro-particles, with

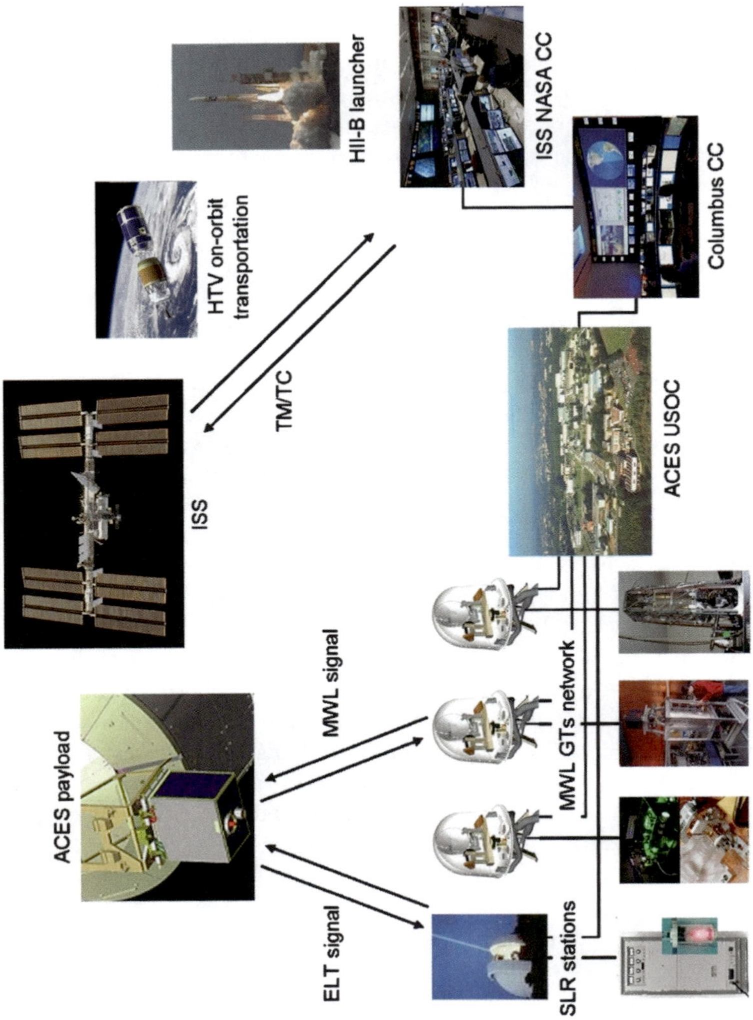

Fig. 3.1 Operational concept for ACES (Source: ESA Erasmus Centre)

a temporal resolution fine enough to investigate the dynamics all the way into the range of e.g. the Einstein frequencies in crystals, thus providing access to a physical regime that was previously not accessible for studies at this level. In the last few years it has been realized that "active" experiments can provide an even bigger and more ambitious scope. The focus now includes the following (remember – all studies at the most basic "*atomistic*" level):

- Fundamental stability principles governing fluid and solid phases.
- Non-equilibrium phase transitions (e.g. electrorheology).
- Phase separation of binary liquids.
- The principles of matter self-organization.
- Concepts of universality at the kinetic level in connection with critical phenomena (i.e. the kinetic origin of renormalization group theory – as developed by Nobel Laureate Kenneth Wilson).
- The physics (structure and dynamics) of cooperative phenomena in "small" nano-systems.
- The kinetic origin of turbulence.
- Non-Newtonian physics effects etc.

So far it has been demonstrated that complex plasmas – with their unique properties of visualization of individual particles and comparatively slow (10^{-2} s) dynamic timescales – can contribute enormously to all these areas of research. On Earth, these studies are complemented by 2-dimensional systems since gravity forces acting on the (comparatively heavy) microparticles lead to flat membrane-like assemblies. 2D studies are of great interest, too, with the result that that this complementarity is very valuable. The tasks ahead will involve utilizing existing and new laboratories on the ISS for dedicated experiments to study these basic strong coupling phenomena, and linking the observations to the complementary 3D research carried out in complex fluid studies. The two fields – complex plasmas and complex fluids – may be thought of as different states of soft matter (relating to the "gaseous" or "plasma" state and "liquids" respectively) with correspondingly different properties (Fig. 3.2).

3.1.1.6 Dust Physics

"Dust" is everywhere! Whilst on Earth it is often regarded as a nuisance, in space possibly even a hazard, it is nevertheless true that without the formation of dust particles in interstellar space we would not exist to lament the existence of dust.

Apart from this obvious question of its origin, it turns out that improved knowledge of "dust physics and chemistry" is generally important – e.g. in the manufacturing of solar cells and electronic circuits, in understanding thunderclouds and lightning, in the physics of the mesosphere, planetary rings and in star and planet formation, to mention just a few important topics. "Dust Physics" aims to provide hard data on some of the most important physical processes [107, 111].

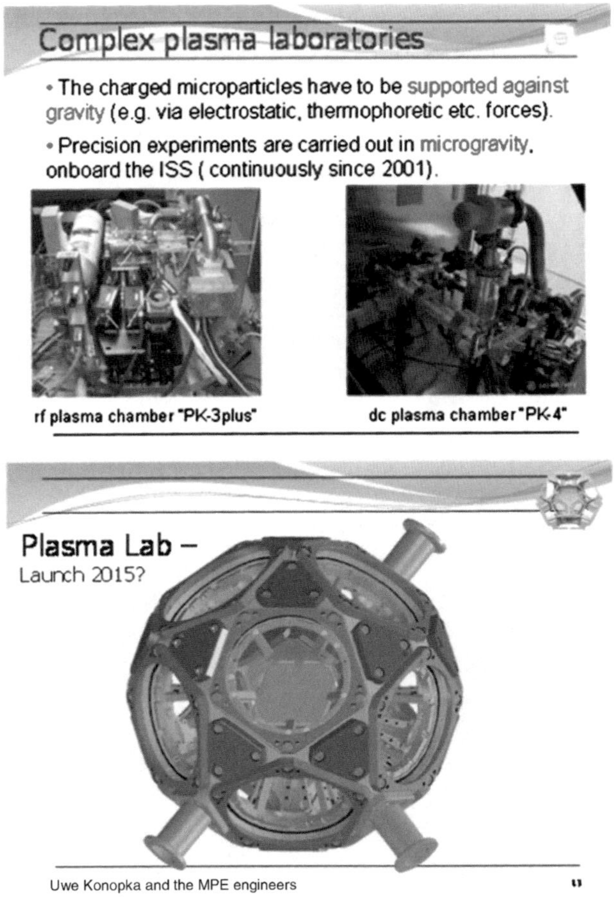

Fig. 3.2 PK-3Plus (which contains an rf plasma chamber) is a joint MPE/JIHT development financed by MPG and DLR. It is currently in operation on the ISS. PK-4 (which contains a dc plasma chamber) is under construction (financed by ESA, MPE and DLR). Launch to the ISS is planned for 2012. Plasma Lab, the next-generation complex plasma experiment, contains a spherical arrangement of electrodes to externally shape and control the binary interaction potential between the microparticles

These are:

- Optical signatures to identify size, shape and chemical composition of the particles.
- Collisions – determining the boundary between adhesive and destructive collisions for a range of velocities, impact parameters and materials.
- Coagulation – the fractal growth of particles under different transport conditions (diffusive, convective), growth rates and structures.
- Condensation of molecules onto dust particles (particularly ice), the transition from epitaxial growth to amorphous shells.

Fig. 3.3 Coagulated dust particle (Source: Walter, TU München)

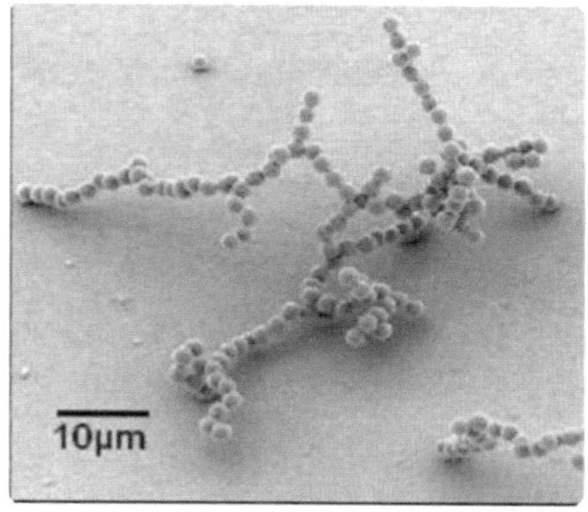

- Triboelectric charging.
- Collisional charge exchange – a process considered to be of major significance in atmospheric lightning.
- Transport and interactions under microgravity.
- Homogeneous nucleation and particle growth, a major process in astrophysics and on Earth in meteorology.
- Dust transport (e.g. photon pressure, the photophoretic effect, dust-gas friction etc.).

In most of the research topics in dust physics, microgravity provides a unique – even essential – environment. For one thing, interstellar, protostellar and planetary ring dust phenomena occur under weightlessness (so it appears reasonable to also use such conditions in experiments) but in addition, some processes require adequate observation time and controlled environments that cannot be achieved in the Earth's gravitational field (Fig. 3.3).

3.1.1.7 Vibration and Granular Matter Physics

At first glance it seems strange for "granular matter" – close-packed assemblies of near-identical and/or size-distributed particles – to be researched in space under microgravity conditions, especially under vibrations. Vibrations are employed to give them energy, that is "temperature" [17, 18, 98, 220]. What is mostly not realized is the enormous scope of granular matter in industry (sand, gravel, grains etc. are the most obvious, with toner particles, colloids, paints etc. at a finer scale) and the surprisingly complex issues involved in size sorting, storage, stability, transport, filling etc. Size sorting can be achieved under gravity e.g. by vibration: the larger particles then migrate to the surface, which may seem counter-intuitive as they are heavier.

In order to study the processes involved in granular matter physics, to understand and model them for the benefit of better and more controlled application on Earth, it is imperative to vary the parameters influencing these processes. One of these parameters, which on Earth is a constant, is gravity. In space, gravity is absent (or very small). This has several benefits for fundamental studies:

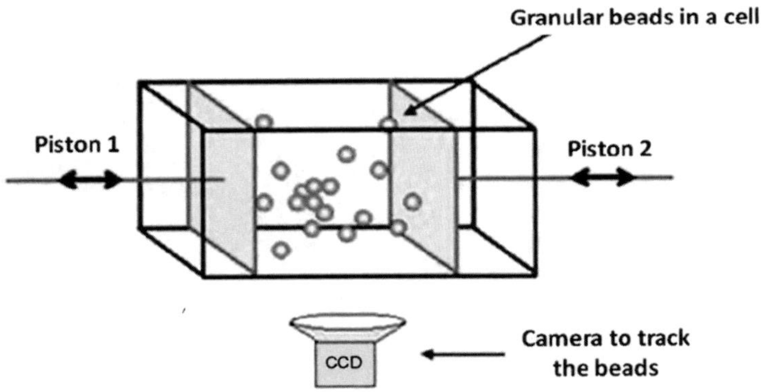

Fig. 3.4 The VIP-Gran concept (Source: ESA Erasmus Centre). The bead vibration will be driven by pistons moving back and forth in a cell volume of 30 × 30 × 70 mm. The cell will be exchangeable. The main diagnostic tool will be a CCD camera

- bigger particles can be used
- time scales for experiments are larger
- the role of fluctuating forces (e.g. vibration) can be studied without "interference" by a macroscopic directed force
- processes can be studied under controlled and variable conditions
- reliable models can be developed that can benefit industrial processes.

Whilst all of this seems very "application oriented" there is also a fundamental aspect to this research. This has to do with self-organization of "hard sphere" matter. In previous examples – complex plasmas and complex fluids – we discussed strongly interacting systems with a soft interaction potential (a Debye–Hückel potential in the case of complex plasmas) on the one hand, and an overdamped hard sphere potential (complex fluids) on the other. Granular matter closes a "systemic gap" – by providing a virtually undamped hard sphere system. In this sense a new regime of parameter space becomes available for studying self-organisation processes (Fig. 3.4).

3.1.2 Fluid Physics by Daniel Beysens

Daniel Beysens

CEA-Grenoble and ESPCI-ParisTech, 10 rue Vauquelin, 75005 Paris, France

Fluids (gas, liquid) are states of matter that are ubiquitous in science and technology, and essential for life. This is why many problems of fluid physics are also encountered in the other fields of this book. The management of fluids in space is a challenge. Their behaviour is indeed markedly different in the space environment than on Earth. Instead of being submitted to the steady Earth gravitational acceleration, fluids in space have to face low gravity and time-dependent acceleration. Forces, such as capillary forces, which are usually small or negligible and are generally ignored on Earth, will become dominant and lead to unexpected and counter-intuitive behaviour. Others, such as buoyancy forces, disappear or are least greatly reduced, causing e.g. diffusion or thermocapillary motion to prevail over convection.

Fluid behaviour is generally described by a non-dimensional number that measures the influence of one quantity on another. The space environment, by reducing the value of the acceleration, will cause the following numbers to change by a large value (this list is not limitative):

Bond number Bo $= \frac{\sigma}{g \Delta \rho L^2}$: capillary force/gravity force with L the typical length of the container, $\Delta\rho$ the gas-liquid density difference, g the earth acceleration constant and σ the gas-liquid surface tension

Capillary number Ca $= \frac{\sigma}{\eta V}$: capillary velocity/fluid velocity, with η the shear (or dynamic) viscosity and V the fluid velocity

Peclet number Pe $= \frac{VL}{D}$: fluid velocity/diffusion, with D the diffusion coefficient (thermal or solutal)

Marangoni number Ma $= \frac{\sigma_T \Delta T L}{\eta D}$: thermocapillary motion/diffusion, with σ_T the surface tension thermal derivative and ΔT the temperature difference

Weber number We $= \frac{\rho L V^2}{\sigma}$ or inertial/capillary energies, with ρ the fluid density.

The investigations in fluid physics aim to predict the new behaviour of fluids in space and explain intriguing observations. The means used for the experimental investigation are Ground-Based Facilities (drop tower, magnetic levitation, density matching of mixtures), parabolic flights with planes and sounding rockets and satellite means (Space Shuttle, Mir, ISS...) that are described above in Chap. 2.

The data that are obtained can be classified as:

- Housekeeping data (temperature, pressure, power consumption, etc.)
- Micro-acceleration data
- Experiment-specific data (video, temperature, pressure, fluid velocity, etc.)
- Time data under different clocks (UT, Mission Elapsed Time, Experiment time, etc.) enabling a precise chronology and coincidence of events to be obtained with the observation and measurement data (video, temperature and pressure sensors, etc.)
- Samples are infrequently produced onboard (interdiffusion measurements, metallic foam...) to be recovered and analysed on the ground.

As usual, some difficulties may arise from the different data formats, supports and standards that were used at different times and by different space agencies (magnetic tapes, video standard, etc.).

In what follows we give an overview of the main fields. More details can be found in the references listed below and in documents [180, 228]. The areas of investigation are concerned with vapour-liquid phase transition, with the particular role of the critical point and heat transfer by boiling, dynamic wetting phenomena with the thermocapillary or Marangoni effect and the stability of liquid bridges, the determination of the diffusion constant in a buoyancy-free environment, the hydrodynamic behaviour of single- or two-phase fluids under external solicitations (accelerations, vibrations) and the behaviour of foam. These research projects are characterized by a strong coupling between theory and numerical simulations, both in the design phase of the experiments and the post-flight analysis.

3.1.2.1 Supercritical Fluids and Critical Point Phenomena

The critical point is the starting point of a new state for gas and liquids when pressure and temperature attain a level such that liquid and gas cannot coexist any more (Fig. 3.5a). They mix together as a dense gas, a gas with the density of a liquid, and are called "supercritical fluids". For carbon dioxide, the critical point occurs at pressure 73 bar and temperature 31°C. For water, the critical point pressure is 220 bar and temperature is 374°C. In the vicinity of the critical point, all fluids behave in a similar manner. Studying one fluid enables the properties of all fluids to be deduced, thanks to so-called "critical point universality", for whose discovery Kenneth G. Wilson won the Nobel Prize in 1982.

The study of pure fluids near their critical point and, to a lesser extent, liquid mixtures near their dissolution critical point, reveals several exceptional features [13]:

- extreme compressibility. As a result, fluids become compressed under their own weight on Earth, even at millimetre scale, density is no longer homogeneous and the critical point cannot be reached.
- extreme thermal dilatation. Minute temperature gradients therefore produce large density gradients. The fluid becomes subjected to strong turbulent convective flows even under extremely small temperature differences.
- in addition, and this is a common feature of binary liquids and fluids, the process of phase transition is mostly governed by buoyancy, the domains of both phases being convected upwards (lower density) or downwards (higher density). Although when nearing the critical point the density difference between both phases approaches zero, the surface tension also approaches zero (but more strongly), with the result that the Bond number eventually tends toward zero: the closer to the critical point, the greater the influence of gravity.

These phenomena become increasingly pronounced as the fluid or the liquid mixture nears its critical point. The high sensitivity of near-critical fluids emphasizes all effects, which cannot be studied without buoyancy effects on earth. This is also the case of the effect of vibrations whose influence is detailed in a special section below.

The weightless environment consequently allows data to be obtained very close to their critical point. This close vicinity has enabled the following investigations to be made.

Testing Universality

Gas-liquid, binary liquids, polymer blends... belong to the same universality class as the Ising system for ferromagnetism transition. A number of important properties such as susceptibility, specific heat, etc. obey scaling laws with universal exponents when asymptotically close to the critical point. This behaviour has been precisely tested under weightlessness (Fig. 3.5b).

Dynamics of Phase Transition

Domains of one phase nucleate and grow at the expense of the other phase. The process, once buoyancy has been suppressed, has been shown to obey universal master laws.

New Process of Thermalization

The "Piston effect" was discovered under microgravity conditions. In a closed sample submitted to a temperature rise or heat flux at a border wall, temperature rises very rapidly due to adiabatic heating by the expansion (heating) or contraction (cooling) of the thermal boundary layer [274]. In contrast to the critical slowing down of thermal diffusion, this process becomes more rapid as the critical point is neared, leading to "critical speeding-up" instead. On Earth, this phenomenon competes with buoyancy flows to accelerate thermal equilibration.

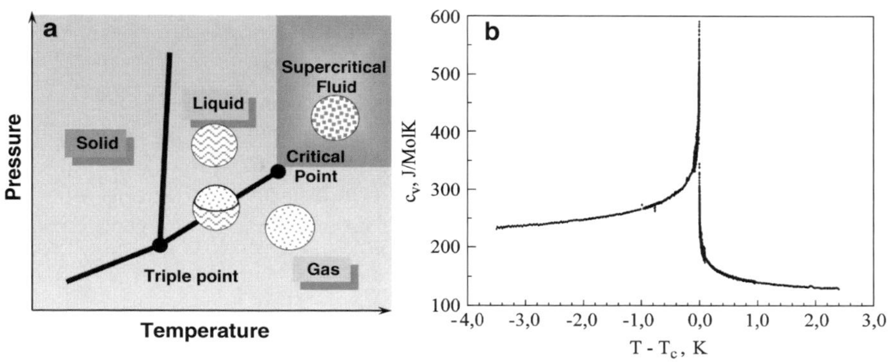

Fig. 3.5 (**a**) Phase diagram of a pure substance in the temperature – pressure plane. The supercritical "state" corresponds to a compressed gas that exhibits the density of a liquid. (**b**) Critical anomaly of the specific heat at constant volume (C_v) measured under zero-gravity in SF_6 (Spacelab D2, 1993) (Source: Haupt and Straub [102])

Supercritical Properties

Fluids such as oxygen and hydrogen are also used in space under supercritical conditions because they show up as a homogeneous fluid irrespective of spacecraft or satellite accelerations – or absence of accelerations – and thus irrespective of the orientation of the gravity vector. In addition, supercritical fluids show very interesting environmental properties; for example, supercritical carbon dioxide is a very powerful solvent of organic matter (while being harmless for health and the environment). It is also possible to burn dangerous waste, like ammunition, in supercritical water very safely and efficiently. Experiments in space concerning these points have just started.

3.1.2.2 Heat Transfer, Boiling, Two-Phase Flow

The heat transfer coefficient α between a wall and a fluid is classically expressed through Newton's law $\alpha = Q/(T_W–T_F)$. Here Q is the power exchanged between the wall and the fluid (per unit surface) and T_W, T_F are the wall and fluid temperatures, respectively.

Heat transfer classically uses convection and phase change (condensation, evaporation). In the latter case, two-phase flow (vapour and liquid) occurs. In space, the absence of buoyancy can considerably modify the performance of heat transfer. Classically, on Earth, the value for α [$Wm^{-2}K^{-1}$] ranges from 100 to 1,200 (free and forced convection), 2,000 to 45,000 (nucleate boiling), 4,000 to 17,000 (filmwise condensation), 30,000 to 140,000 (dropwise condensation). In space, α is generally found to be lower.

The investigations in low gravity are concerned with the mechanisms involved in the process: evaporation, condensation, boiling and two-phase fluid flow.

Two-Phase Flows

These have mainly been studied in a configuration close to an industrial process by using a two-phase loop experiment, with capillary pumping [148]. Capillary forces become dominant in space and enable to efficiently pump the liquid in a porous medium, preventing the use of a mechanical device. The basic mechanisms (convection in an evaporating phase, drop evaporation, evaporation in a porous medium) are addressed, together with more technical investigations, as the performance of a two-phase loop.

Boiling and Boiling Crisis

Boiling is a highly efficient way to transfer heat. Boiling on earth has been studied extensively by experiment for common fluids and conventional regimes, for

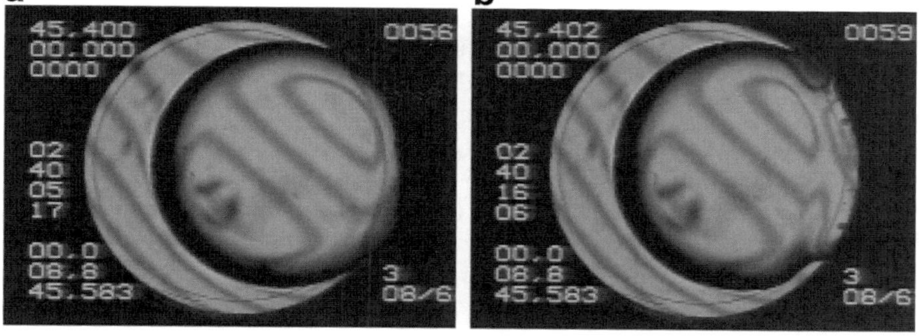

Fig. 3.6 Boiling and bubble spreading under zero-gravity (SF_6, MIR, 1999) (**a**) t = 0, no heat flux at the wall (**b**) t = 11 s under heat flux, vapour spreads at the contact line location due to the recoil force (Source: Garrabos et al. [85])

instance for water at atmospheric pressure and moderate heat flux supplied to the fluid. The basic theory of boiling remains “terra incognita”, in particular, the phenomena very close to the heating surface, at a scale much smaller than the vapour bubbles. The efficiency of industrial heat exchangers increases with the heat flux. However, there is a limit called Critical Heat Flux. It corresponds to a transition from nucleate boiling (boiling in its usual sense) to film boiling where the heater is covered by a quasi-continuous vapour film and the evaporation occurs at the gas-liquid interface. Since the gas conducts heat much less than the liquid, the heat transfer efficiency drops sharply during this transition and the heater heats up, which may cause damage to it if the power is not cut immediately. This transition is called “burnout”, “departure from nucleate boiling” or “Boiling Crisis”.

Starting from low gravity investigations (Fig. 3.6), a vapour recoil mechanism for the Boiling Crisis has been proposed. A fluid molecule leaving the liquid interface causes a recoil force analogous to that created by the gas emitted by a rocket engine. It pushes the interface towards the liquid side in the normal direction. At high enough heat flux, a growing bubble can forcefully push the liquid entirely away from the heating element. The evaporation is particularly strong in the vicinity of the contact line of a bubble, inside the superheated layer of the liquid [189].

3.1.2.3 Interfaces

Interfaces play a key role in many areas of science and technology. One can cite phase transition (evaporation, boiling, solidification, crystallization), combustion, foam and thin film drainage, thermocapillary motion, rheology of suspensions, emulsion stability, all of which are domains that benefit of the low gravity environment.

Liquid Bridges

A liquid bridge is a volume of liquid that is surrounded by another fluid and is attached to more than one solid wall. It differs from a sessile drop (attached to one solid wall) or a free drop. The most simple situation is a liquid bridge spanning the gap between two flat surfaces (more complicated situations occur, e.g. in porous media). The pinning conditions of the liquid-solid-gas triple line is a key parameter.

The most studied bridge [170] is the (standard) cylindrical bridge spanning the gap between two coaxial solid disks. Many studies have focused on statics (equilibrium shape and stability) and dynamics. In this latter area, one must consider the behaviour without thermal effects (disk vibration, rotation, stretching), with thermal gradients (thermocapillary Marangoni flows), diffusion of species (solutal Marangoni flows), phase change (unidirectional solidification – floating zone process), electric and magnetic effects (shape stabilization, convection suppression) and reactive processes (cylindrical flames).

Marangoni Thermo-Soluto-Capillary Flows

Capillary (Marangoni) flows develop when the surface tension varies along the liquid-gas interface, from the low surface tension region to the high surface tension region [39]. The gradient of surface tension induces a surface flow that tends to drag the underlying bulk liquid with it. The gradient can be induced by a temperature difference, inducing a thermocapillary motion proportional to $d\sigma/dT$, or a difference in concentration c of a surface-active species. In this case the motion is proportional to $d\sigma/dc$. The resulting flow velocity can be large, increasing or diminishing the buoyancy (gravity)-induced flows on Earth. When liquid droplets or gas bubbles are submitted to a temperature or concentration gradient, the surface flow makes the drop or bubble move, often rather rapidly.

Such Marangoni flows appear naturally during phase transition where bubbles or drops migrate towards the hottest wall (if $d\sigma/dT > 0$) and convective flows appear on a solid-liquid interface. This is why most of the studies have been performed in configurations close to those encountered in Materials Sciences e.g. in the molten zone crystallisation process (liquid bridge). There is indeed a direct link between the quality of crystals and the flow properties in the liquid phase. Other investigations are classically concerned with an open tank where the free surface is submitted to a temperature gradient between two parallel rigid walls.

Interfacial Transport

The dynamics aspects of adsorption of soluble surface active species (surfactants) benefit from a low-gravity environment [198]. Real-time measurements of liquid-liquid and gas-liquid surface tension enable a better understanding of the very nature of the interfacial transport process, in particular the diffusion in the bulk of the fluid towards the interface, the exchange of matter and the dilatational rheology of the interface.

Foams

Liquid foam (Fig. 3.7) exhibits a cellular structure made up of a liquid surrounded by gas bubbles. Wherever two bubbles press tightly together, the liquid is in the form of a thin film. The foam is not entirely static – unless it has been solidified as e.g. polystyrene or metallic foams. As long as the liquid component is present, the foam evolves under the action of the three following processes: (i) Drainage: the motion of liquid through the foam by gravity. An equilibrium with height is reached, with "dry" foam above and "wet" foam (more than 15% liquid fraction), near the underlying liquid. (ii) Coarsening: the increase with time of the droplet size. This process is due to the diffusion of gas through the thin film as induced by the pressure difference between gas and liquid. Smaller droplets (higher pressure) are thus eliminated in favour of the larger ones (lower pressure). (iii) Rupture: coarsening ends with the rupture of thin films, which causes the foam to collapse.

Low gravity experiments have led to greater understanding of the properties of wet foam properties [217]. Some of these experiments also addressed the quite difficult technology process of the fabrication of metallic foams without the additives that weaken the materials produced on Earth.

3.1.2.4 Emulsions

If shaken vigorously, immiscible liquids, like oil and vinegar or oil and water, form a dispersion of small droplets of one liquid in the other phase, i.e. an emulsion. In general, this emulsion is not stable, and sooner or later the emulsion evolves to the two initial distinct liquid phases. However, some emulsions, like mayonnaise, are stable because the droplets are laid apart by surfactant molecules (here proteins) that migrate to their surface. The control of the stability of emulsions is one of the most important problems in emulsion science and technology.

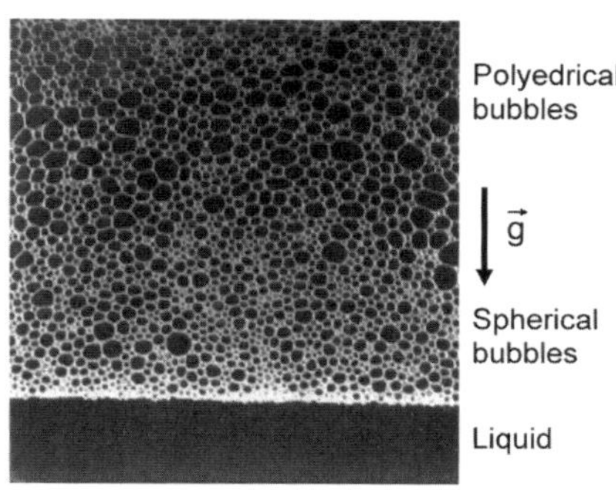

Fig. 3.7 Drainage of a 'wet' foam on Earth. This process does not occur in space where, instead, the wetting front obeys a diffusive process (Source: H. Caps, GRASP)

The main factors that lead to the destabilization of emulsion are the following. (i) Aggregation. Different droplets of the dispersed phase aggregate in clusters. (ii) Coalescence. Two or more droplets in contact fuse together. (iii) Ostwald ripening. The liquid in a small droplet diffuses to a neighbouring larger droplet due to the pressure difference corresponding to the different radius of curvature. Although Brownian motion is effective for bringing very small droplets in contact with each other, it is mainly the gravitational forces that eventually "cream" to the surface or "settle" to the bottom.

Experiments in microgravity eliminate the influence of gravitational forces and highlight the other causes of destabilization. Researchers can then investigate the dynamics of surfactant adsorption at the interface, drop-drop interactions and the dynamics of phase inversion in model emulsions (from oil in water to water in oil). Experiments with metallic emulsions have shown that other processes than those described above, such as Marangoni effects, can lead to the destabilisation of emulsions [176].

Giant Fluctuations of Dissolving Interfaces

It was discovered recently that, unexpectedly, large spatial fluctuations in concentration can take place during a free diffusion process, such as the process occurring at the interface of liquids undergoing a mixing process [254]. Such fluctuations of concentration (and density) are due to a coupling between velocity and concentration fluctuations in the non-equilibrium state. As the amplitude of the fluctuations is limited by gravity, experiments indeed observed the fluctuations increase ('giant' fluctuations) when gravity was not present (Fig. 3.8). The experiment was performed at the interface between a liquid and its vapour above the critical point (see above) where liquid dissolves in the vapour. The observation results of such giant fluctuations influence other types of microgravity research, such as the growth of crystals.

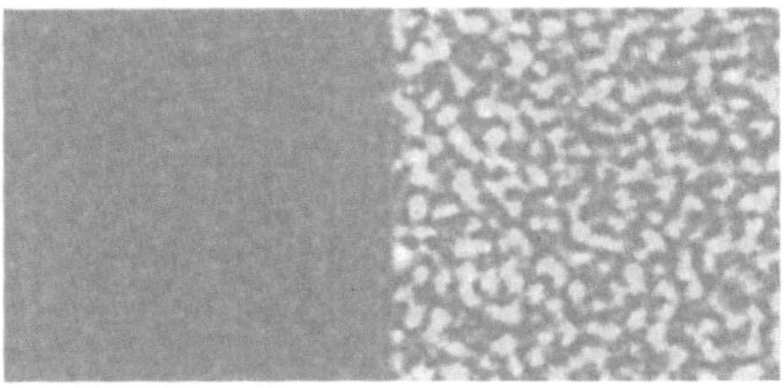

Fig. 3.8 Fluid theory confirmed during a Foton flight. False colour shadowgraph images (5 mm × 5 mm) from polystyrene in toluene, showing temperature fluctuations on Earth (*left*) and aboard Foton-M3 (*right*). The fluctuations in normal gravity are barely visible (Source: ESA)

3.1.2.5 Measurements of Diffusion Properties

As described by Fick's law, diffusion is the tendency of species to spread uniformly in solutions. The diffusion coefficient D is the ratio between the mass flow flux $\overrightarrow{J}$ of a component and the concentration difference between two points in the system. For a binary liquid solution, this law is $\overrightarrow{J} = -D\overrightarrow{\nabla}c$, with $\overrightarrow{\nabla}c$ the concentration gradient, and a component always diffuses from high to low concentration points. It is a slow process, e.g. it takes 10^4–10^5 s to homogenize a solution over 1 cm length. In the presence of a temperature gradient, thermodiffusion – the "Soret effect" – takes place, with the same typical times as above.

These long characteristic times mean that slow convection, as encountered in Earth-bound measurements, can seriously perturb the concentration field [253]. Low gravity measurements eliminate such disturbing motions inside non-homogeneous samples. (Note that the environment has to be of high quality. Vibrations, as discussed below, can alter the quality of data). Many experiments have been conducted in materials of scientific or industrial interest, such as molten salts, metallic alloys and organic mixtures. They led to values that are significantly lower than those measured on Earth. These values help to discriminate between the many different theories that aim to predict diffusion behaviour, some of which are quite difficult to formulate in cases of complex systems. Reliable data have also been obtained for the computer industry (solidification, crystallisation of materials) and oil companies (diffusion coefficient of crude oils).

3.1.2.6 Vibrational and Transient Effects

Most experiments that are performed under space microgravity conditions are selected because of their sensitivity to gravity effects. They are thus also sensitive to acceleration variations that correspond to manoeuvres - leading sometimes to unwanted sloshing motions - and, and to erratic or non-erratic vibrations ("g-jitters").

Transient and Sloshing Motions

During the cutoff or reignition phase of a spacecraft engine, or during the orbiting manoeuvres of a satellite, the motion of two-phase fluids (e.g. liquid oxygen or hydrogen in equilibrium with their vapour) in the reservoirs can exhibit severe sloshing motion that can even lead to an interruption of fuel flow to the engines.

The physics of the problem is complicated by the fact that the free surface of the liquid is not simply flat. If the fluids and the solid mass are comparable in magnitude, the dynamics of the system are coupled [260]. Experiments to validate three-dimensional Computational Fluid Dynamics simulation have been carried out mostly on model fluids by using scaling with the Bond and Weber numbers, which are the main numbers involved in the problem. Only a few experiments have been performed in real spacecraft tanks or with real cryogenic fluids (oxygen, hydrogen).

Vibrational Effects

Knowledge, prediction and minimization of vibration effects are therefore a necessity when dealing with the control of space experiments. One could also use the natural mechanical noise of the spacecraft to reduce or improve the technical characteristics of the experimental set-up, just by positioning the set-up and orientating it carefully in space.

Vibrations generally tend to direct inhomogeneities parallel or perpendicular to the acceleration direction, depending on the amplitude and frequency. Density inhomogeneities can be induced in a simple fluid by temperature gradients, or created by the coexistence of two phases (liquid and vapour). Vibrations applied to mechanical systems can induce destabilization or stabilization, depending on the characteristic features of the vibrations (frequency, amplitude) and the direction of vibration with respect to that of the density gradient.

Vibrations can easily provoke average motions in fluids with density inhomogeneities, counterbalancing or emphasizing the gravitational flows on Earth and inducing effects in space that are similar to those provoked by gravity. In particular, thermal instabilities similar to those encountered on Earth, such as the well-known Rayleigh-Bénard instabilities, can be induced by vibrations [86]. Vibrations can therefore be of interest for the management of fluids and can be considered as a way to create an "artificial" gravity in space.

A number of investigations have been conducted in space during phase transition, especially near a vapour liquid critical point (Fig. 3.9a). Thermo-vibrational aspects have been the object of experiments with liquid (Fig. 3.9b) and liquid solutions to evaluate the effect of vibration on the Soret (thermodiffusion) coefficients and in homogeneous, supercritical fluids near their critical point, where the effects are magnified. Experiments have also been conducted in materials science (solidification). They all conclude on the significant effect that vibrations can induce on fluid behaviour, sometimes being able to act as artificial gravity.

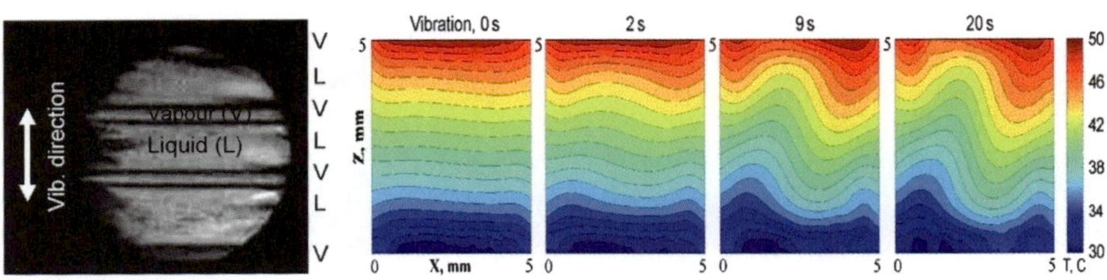

Fig. 3.9 (**a**) Sample of liquid CO_2 in presence of its vapour in a Maxus 5 sounding rocket when submitted to vibration of 300 μm amplitude and 20 Hz frequency. After a temperature change, liquid droplets and vapour bubbles order into periodic well-defined liquid and vapour phases with plane interfaces similar to Earth-bound conditions. (Source: Beysens et al. [17]). (**b**) The evolution of temperature field in a vibrated square sample of 5 mm length filled with alcohol (*side view*) during parabola weightlessness conditions. Vibration: 45 mm amplitude, 4 Hz frequency. Well-organised rolling manoeuvres provide experimental evidence of thermal vibrational convection and verify existing theoretical studies (Source: Mialdun et al [174])

3.1.2.7 Biofluids: The Microfluidics of Biological Materials

Although the basic constituents of a biological fluid, such as blood, are of the order of μm, and thus weakly gravity-dependent, the absence of gravity can affect the behaviour of biofluids [161]. The other dimensional values (diameter and length of the vessel) are indeed large. Investigations in weightlessness of vesicles, which are good models for blood cells, show that vesicles can undergo temporal oscillations under the influence of shear flow. These studies also have industry relevance, as micron-sized particles that are very similar to blood compounds in size (red blood cells, white blood cells) can be separated by a hydrodynamic focusing method (the so-called "split" technique). Microgravity is essential for improving the process. In particular, it was shown that transversal migration of cells, which is detrimental to separation efficiency, is due to shear-induced hydrodynamic diffusion.

3.1.3 Combustion by Christian Chauveau

Christian Chauveau

CNRS – INSIS – ICARE, 1C, avenue de la Recherche Scientifique,
45071 Orléans Cedex 02, France

3.1.3.1 Introduction

Combustion is a rapid, self-sustaining chemical reaction that releases a significant amount of heat and as such involves elements of chemical kinetics, transport processes, thermodynamics and fluid mechanics. Combustion is a very common phenomenon, which has accompanied humanity throughout its history, whether as campfires, candle flames, domestic burners for heating or cooking, or the burning of hydrocarbon fuels in internal combustion engines. For combustion to occur, three components are required: a fuel, an oxidizer, and a heat source. Fuels can be solid, liquid or gas. Paper, wood and coal are examples of solid fuels, but metallic fuels such as iron, aluminium or magnesium must also be considered. Examples of liquid fuels include gasoline and kerosene. Butane, propane, methane and hydrogen are examples of gaseous fuels. Oxidizers can also be found in all three states, but the most common is the gaseous oxygen present in air. The combustion reaction is generally initiated with a heat source: for example, by friction heating, electrical cable heating, a spark, etc.

Combustion is a key element in many technological applications in modern society. In itself, combustion is one of the most important processes in the world economy. Combustion underlies almost all systems of energy generation, domestic heating, and transportation propulsion. It also plays a major role at all stages in the industrial transformation of matter, ranging from the production of raw materials to the complex assembly of industrial products. Although combustion is essential to our current way of life, it poses great challenges to society's ability to maintain a healthy environment and to preserve vital resources for future generations. Improved understanding of combustion will help us to deal more effectively with the problems of pollution, global warming, fires, accidental explosions and the incineration of dangerous waste. In spite of more than a century of extensive scientific research, many fundamental aspects of combustion are still poorly understood.

The objectives of scientific research into microgravity combustion are initially to increase our knowledge of the fundamental combustion phenomena that are affected by gravity, then to use the research results to advance science and technology related to combustion in terrestrial applications, and finally to tackle questions of security related to fires on board spacecraft. Microgravity combustion scientists undertake experiments both in ground-based microgravity facilities and orbiting laboratories and study how flames behave under microgravity conditions.

Microgravity research allows new experiments to be carried out in which buoyancy-induced flows and sedimentation are virtually eliminated. Combustion usually involves large temperature increases resulting in a consequent reduction in density, ranging from a factor of two to ten depending on the situation. As a result of this density change, the combustion processes in normal gravity are usually strongly influenced by natural convection. The rise of hot gas creates a buoyancy-induced flow favouring gas mixing from the fuel, oxidizer and combustion products. Under conditions of reduced gravity, natural convection is cancelled (or greatly reduced), and therefore the characteristics of combustion processes can be profoundly altered.

The reduction of buoyancy-induced flows has several features that are particularly useful for fundamental and applied scientific research on combustion. By eliminating the effects of natural convection, a quiescent environment is created, conducive to more symmetrical results. This facilitates comparisons with numerical modelling results and with theories. Furthermore, the elimination or drastic reduction of buoyancy-induced flows can reveal and highlight weaker forces and flows that are normally masked, such as electrostatics, thermo-capillarity and diffusion. Lastly, the elimination of disturbances caused by buoyancy forces can increase the duration of experiments, thus allowing the examination of phenomena over longer time scales.

For purposes of simplification, the numerical models developed in combustion research often assumed that the mixture of the initial components is homogeneous. Sedimentation affects combustion experiments involving drops or particles, since the components with the highest density will be driven down into the gas or liquid, and hence their movement relative to other particles creates an asymmetric flow around the falling particle. The presence of these concentration gradients in the mixture before combustion complicates the interpretation of experimental results. In normal gravity conditions, experimenters must implement devices to stabilize and homogenize dispersed media, e.g. supports, levitators or stirring devices. In microgravity, gravitational settling is almost eliminated, allowing the stabilization of free droplets, particles, bubbles, fog, and droplet networks for fundamental studies on ignition and combustion in heterogeneous media.

To date, scientific research in combustion has shown major differences in the structure of different flames burning either in microgravity or under normal gravity. Besides the practical implications of these results in terms of combustion efficiency, pollutant control, and flammability, these studies have established that a better understanding of the sub-mechanisms involved in the overall combustion process is possible by comparing the results obtained in microgravity with those obtained in normal gravity.

While microgravity is the operational environment associated with Earth-orbiting space laboratories, it is important to note that "ground-based facilities" allowing gravity reduction also serve the scientific community and enable relevant combustion studies to be carried out. Experiments conducted in suborbital sounding rockets, during the parabolic trajectory of an aircraft-laboratory, and free-fall drop towers, significantly complement the limited testing opportunities available aboard the International Space Station. In fact, the contributions of research conducted in these so-called "ground-based" microgravity facilities have been essential to the acknowledged success of microgravity combustion research. These helpful facilities allow us to consider microgravity as a tool for combustion research, in the same way as an experimenter can vary pressure or temperature; microgravity can also act on the gravitational acceleration parameter. Additional contributions of high value to microgravity combustion research come from "normal-gravity" reference ground tests and from analytical modelling.

The reader will find in this paper a non-exhaustive review of the different research areas in which experiments have been conducted in microgravity. This document is not intended to be a complete bibliography of every investigation of combustion in microgravity; the reader is encouraged to consult review articles [78, 145, 216, 269].

3.1.3.2 Premixed Gas Flames

The most spectacular advances in the field of microgravity combustion occurred with studies on premixed flames. In premixed gas flame research, the fuel and oxidant gases are thoroughly mixed before ignition. Scientists are interested in flame velocity as a function of both the type of fuel and oxidizer used and the concentration ratio of the two components. The mixture is considered flammable if propagation initiated by a spark is successful, which only occurs if the mole fraction of fuel in the mixture exceeds a minimum value, called the lean flammability limit, and if it is less than a specified maximum value (the rich flammability limit), above which the oxidizer quantity is insufficient to sustain propagation. These lower and upper flammability limits are of considerable interest in terms of both safety and fundamental science. Most of these phenomena have been extensively studied in laboratories but always under the influence of normal gravity. Some aspects of the combustion process are known to be masked or modified by the presence of gravity. For example, it is difficult to measure laminar flame speeds in the presence of buoyant convection under conditions close to the flammability limits, and numerical models usually neglect gravity in order to focus on chemical kinetics.

In addition, premixed gas flames can develop different instabilities driven by hydrodynamic, thermal diffusion, kinetic, or chemical mechanisms. These effects, which are often weak in normal gravity, can then be isolated and identified by microgravity experiments. As an example of a spectacular result, stationary premixed spherical flames (i.e. flame balls), whose existence was predicted by theory but had never been confirmed by any experiments in normal gravity, were observed uniquely in microgravity [215]. Thus, a number of unique phenomena in the case of premixed flames occur under microgravity conditions.

3.1.3.3 Gaseous Diffusion Flames

In this type of flame, the fuel and oxidant gases are initially separated. They then diffuse into each other, thus feeding a reaction zone conducive to flame stabilization. Diffusion flames also behave very differently under reduced gravity conditions due to the reduced influence of buoyancy forces. Unlike premixed flames, diffusion flames do not propagate and have fewer instability mechanisms.

Diffusion flames, from a gaseous fuel jet injected into air, observed in microgravity experiments have revealed features different from those of their counterparts under normal gravity: more spherical or more symmetrical geometry, greater flame thickness and stability, lower temperature and, more significantly, higher soot production [126, 150, 251]. Again, these observations are essential for the fundamental understanding of phenomena, and are used to develop models with potential applications in creating effective strategies to control soot formation in many practical applications.

3.1.3.4 Liquid Fuel Droplets and Sprays

In most applications that use liquid fuels, the fuel is injected in the form of tiny droplets, thereby increasing its specific surface area and thus promoting its vaporization and combustion. Understanding the complete process of fuel droplet combustion is therefore very important for practical applications. From a scientific point of view, the burning of an isolated, spherically symmetric droplet is the simplest example of non-premixed combustion, which involves the participation of a liquid phase (fuel) in the gas-phase diffusion flame. Research on the combustion of isolated liquid droplets of pure fuel makes it possible to study the interactions of physical and chemical processes in a simplified and idealized geometry. In real applications, since the diameters of the injected droplets are small (in the order of tens of microns), the effects of buoyancy forces are consequently low. On the other hand, the times of vaporization or combustion associated with these small droplets are also extremely short, thus complicating experimental measurements. The spherically symmetric burning configuration obtained in microgravity allows scientists to develop detailed theoretical models with a simplified one-dimensional representation. Moreover, the absence of disturbance makes it possible to conduct microgravity experiments with larger droplets, and therefore with a longer observation time, allowing a detailed study of both transient and quasi-steady phenomena in the liquid phase and in the gas phase, as well as flame extinction phenomena. The literature abounds with studies of droplets in microgravity, since the pioneering work of Kumagai [137], and extensive research has been conducted around the world. These studies have identified numerous unique phenomena from a fundamental point of view, but they have also led to the development of many experimental correlations that are useful in applications. These correlations involve the parameters that directly influence the droplet evaporation rate, e.g. temperature, pressure, composition of the gaseous environment, composition of the liquid fuel, either pure or multicomponent, and also the experimental constraints, e.g. supported or free-floating droplets, residual gravitational acceleration, etc. [40, 41]

More recent studies examine interaction phenomena, and implement several drops distributed either linearly or in complex three-dimensional networks or in the ultimate form of aerosol or sprays. Ignition and flame propagation in these structures are closely scrutinized. Increased knowledge of the heterogeneous combustion process resulting from these studies should lead to major improvements in the design of combustors using liquid fuels [175, 193].

3.1.3.5 Fuel Particles and Dust Clouds

The field of combustion research has important implications for practical processes and in the prevention of fire and explosion. The combustion processes of metals are of considerable interest for applications as diverse as solid rocket propulsion systems, oxygen handling, synthetic ceramics, and metal cutting. Clouds of coal dust have the potential to cause mine explosions and grain-dust clouds can cause silos and grain elevators to explode. It is particularly difficult to study the fundamental combustion characteristics of fuel dust clouds under normal gravity because initially well dispersed dust clouds quickly settle due to density differences between the particles and the surrounding gas. This sedimentation process induces a stratification forming non-uniform fuel-air ratios throughout the cloud. Experiments conducted in microgravity allow fuel-dust clouds to remain evenly mixed [30].

In the metal-particles combustion studies, the application of advanced diagnostics reveals novel features of metal combustion. Pulsating combustion phenomena, due to the formation and the collapse of the oxide layer, have been demonstrated for the first time [60, 149]. Due to the very high temperatures encountered in these experiments, there is an increase in the effects of buoyancy forces, making experiments under reduced gravity necessary.

3.1.3.6 Flame Spread Along Surfaces

The inhibition of flame spreading along both solid and liquid surfaces is of primary importance in fire safety. Flame spread involves the reaction between an oxidizer gas and the vapour from the pyrolysis of a condensed-phase fuel. Experimental studies have revealed major differences between normal and reduced gravity conditions, concerning the ignition and flame spreading characteristics of solid and liquid fuels. Many configurations with solid fuels, e.g. PMMA, or liquid pools, e.g. 1-butanol, have been investigated in microgravity conditions, with different configurations of the gas phase, either stagnant or with a low forced convection, and different oxygen concentrations [79, 221]. Such data could enhance our understanding of these effects and provide "benchmarks" for numerical models. As it is not possible to test both flame spread and material flammability of all materials on board spacecraft, numerical models can provide this knowledge bridge between environments at different gravity levels. The knowledge gained from these studies may also lead to a better understanding of dangerous combustion reactions on Earth. Microgravity experiments eliminate the complexities associated with buoyancy effects, providing a more fundamental scenario for the development of flame-spreading theories.

3.1.3.7 Smouldering Combustion

Smouldering is a flameless combustion process that takes place in porous solid fuels, and is characterized by a heterogeneous surface reaction that spreads inside the fuel material. A well-known example of smouldering combustion is burning cigarettes. Smouldering is important both as a fundamental combustion mechanism and as a fire precursor, since fires are often triggered by the transition from slow smouldering to rapid blaze. When a porous fuel smoulders for a long period of time, it can create a large volume of vaporized fuel, which may react suddenly if an additional oxidizer supply occurs. This transition to deflagration often results in catastrophic fires. As the heat is produced very slowly in this process, the burning rate is quite sensitive to heat transfer, and thus natural convection effects are particularly important. Transport and reaction processes in smouldering combustion are complex, and the removal of gravity substantially simplifies their study [10, 264].

3.1.3.8 Diagnostics and Apparatus Development for Combustion Science

The scientific returns from microgravity combustion experiments are directly related to the capabilities of quantitative measurement of significant variables. The instrumentation and measurement techniques available in laboratories must be implemented and adapted to the unique and severe operating conditions of microgravity. A better understanding of chemical kinetics mechanisms in combustion experiments requires an accurate and nonintrusive measurement of various quantities such as the flow velocity, temperature, species concentration, the volume fraction and size of soot particles. The development of diagnostic techniques is therefore based on the evolution of those used in the laboratory at normal gravity, and integrates them into modules designed for use in space by adapting them to the demands of the latter. The latest laser-based equipment can be customized and installed onboard, for Particle Image Velocimetry (PIV) for velocity measurement, interferometry for density, Rayleigh scattering for temperature, spectroscopic measurements for species, or Laser Induced Incandescence (LII) for soot formation among other examples [150, 194].

While most microgravity combustion experiments have been conducted using dedicated and unique experimental apparatus, the recent commissioning of the CIR combustion module in the Fluid Combustion Facility (FCF) experimental rack aboard the ISS should enable more investigators to have access to this microgravity environment. The Combustion Integrated Rack (CIR) features a 100-litre combustion chamber surrounded by optical and other diagnostic packages including a gas chromatograph. Experiments are conducted in the chamber by remote control from the Telescience Support Center (TSC). The CIR is the only rack on the ISS dedicated to combustion experimentation. The CIR provides up to 90% of the required hardware to perform a majority of future microgravity combustion experiments on board the ISS. The remaining 10% of hardware will be provided by the principal investigator hardware development teams.

3.1.3.9 Conclusions

Compared to experimental combustion studies in laboratories, the number of microgravity experiments is small. Nevertheless important discoveries have already emerged from microgravity combustion investigations. The numerous facilities existing now, both "ground-based" microgravity facilities and on board the ISS, made available to the scientific community by space agencies, suggest that new microgravity combustion experiments will significantly advance fundamental understanding in combustion science. It is hoped that this will help to maintain a healthy environment and preserve vital resources for future generations.

3.1.4 Materials Science

Hans Fecht

University of Ulm, Institute of Micro and Nanomaterials, Albert-Einstein-Allee 47, 89081 Ulm, Germany

3.1.4.1 Introduction

Materials science is an interdisciplinary field dealing with the properties of matter and its applications to various areas of science and engineering. This science basically investigates the relationship between the structure of materials and their various properties. It includes elements of applied physics and chemistry, as well as chemical, mechanical and electrical engineering. With significant attention to nanoscience and nanotechnology in recent years, materials science has been propelled to the forefront.

The research – from a fundamental and applied point of view – is concerned with the synthesis, atomic structure, chemical element distribution and various favourable properties of materials and structures. The basic understanding and optimisation of properties and structures is further supported by sophisticated computer models from the nano – to the macro-scale and leads to a wide range of applications in industrial products. Prominent examples are found in the aerospace, automotive, biomedical, energy and microelectronics industries.

Material scientists originally devoted most of their efforts in studying the final products, i.e., the solid state of materials, their microstructure and their mechanical and thermal properties. However, in the last 10–20 years and based on new experimental techniques, a change in paradigm has taken place, and the importance of the liquid phase has been recognized [80]. In this regard, it is interesting to note that almost 100% of all metallic products are, at some stage, produced through solidification and casting processes. Consequently, this field of new materials, processes and products constitutes a major backbone to industries worldwide. A recent estimate indicates that the millions of tons of castings produced worldwide are worth approximately 100 billion Euros per year [74].

Solidification from the melt leaves its fingerprints in the final material, and hence it is of outmost importance to understand the properties of the molten state and its solidification behaviour. The prominent feature of fluids, namely their ability to flow and to form free surfaces, poses the main difficulty and challenge in their theoretical description. The physics of fluids is governed by the Navier-Stokes equation and by the ubiquitous presence of convection. In addition, when dealing with metallic materials, the high temperatures necessarily involved lead to experimental difficulties, the most trivial, but also most fundamental, being the suitability of available containers.

Besides the atomic scale inherent to condensed matter and the intermediate scales associated with solidification microstructures, fluid flow driven by gravity generally occurs in the melt at the macroscopic level so that the relevant length scales in casting cover a broad spread from atomic size (capillary length, crystalline defects such as dislocations, attachment of atoms, etc.) to the metre size of the ingot (fluid flow, spacing of dendrite side branches, etc.).

Accordingly, to produce materials that meet ever-higher specific requirements and performance the solidification processing of structural and functional materials must be controlled with ever-increasing precision. It can be foreseen that tomorrow's materials will be optimised in their design and underlie more efficient production conditions, availability of scarce resources and cleaner processes.

The interactive feedback between experiments and sophisticated computer simulations developed within the last 10 years that now drives the design and processing of materials is achieving performances never seen in the past. Thus it becomes possible to control and optimise the defect and grain structure in critical areas of components. In this regard, two major aspects are most essential for the continued improvement of materials processing, with increasing requirements on composition, microstructure and service achievements, which often implies the breaking of technology barriers:

(i) the reliable determination of the **thermophysical properties** of metallic melts for industrial process design in order to understand the fundamentals of complex melts and their influence on the nucleation and growth of ordered phases, together with
(ii) the reliable determination of the formation and selection mechanisms at microstructure scales in order to develop **new materials, products and processes**.

On this basis, "Materials Science" is closely related to "Fluid Physics" and to some extent to "Fundamental Physics", for example in the areas of "phase transformations" and "soft matter". The motivation for performing benchmark experiments in the microgravity environment is straightforward. First of all, in space it is possible to suppress the gravity-induced effects of fluid flow and more subtle sedimentation effects during solidification. Therefore, the contribution of diffusion to mass and heat transport in the melt can be investigated without the complications of buoyancy-driven thermo-solutal convection and sedimentation/flotation. As a result, fresh insights into alloy solidification/processing can be gained with the potential of producing novel materials and structures, i.e., materials processed and designed in space.

3.1.4.2 Scientific Challenges

Casting is a non-equilibrium process by which a liquid alloy is solidified. The liquid-solid transition is driven by the departure from thermodynamic equilibrium where no change can occur. From the standpoint of physics, casting thus belongs to the vast realm of out-of-equilibrium systems, which means that, rather than growing evenly in space and smoothly in time, the solid phase prefers to form a diversity of microstructures [19] (Fig. 3.10).

Actually, the relevant length scales in casting are spread widely over 10 orders of magnitude. At the nanometre scale, atomic processes determine the growth kinetics and the solid-liquid interfacial energy, and crystalline defects such as dislocations, grain boundaries and voids are generally observed. Macroscopic fluid flow driven by gravity or imposed by a stimulus (electromagnetic field, vibration, etc.) occurs in the melt at the metre scale of the cast product. The characteristic scales associated with solidification microstructures are mesoscopic, i.e., intermediate, ranging from the dendrite tip/arm scale (1–100 μm) to grain size (mm-cm). It follows that the optimisation of the grain structure of the product and inner microstructure of the grain(s) during the liquid-to-solid phase transition is paramount for the quality and reliability of castings, as well as for the tailoring of new advanced materials for specific technological applications [72].

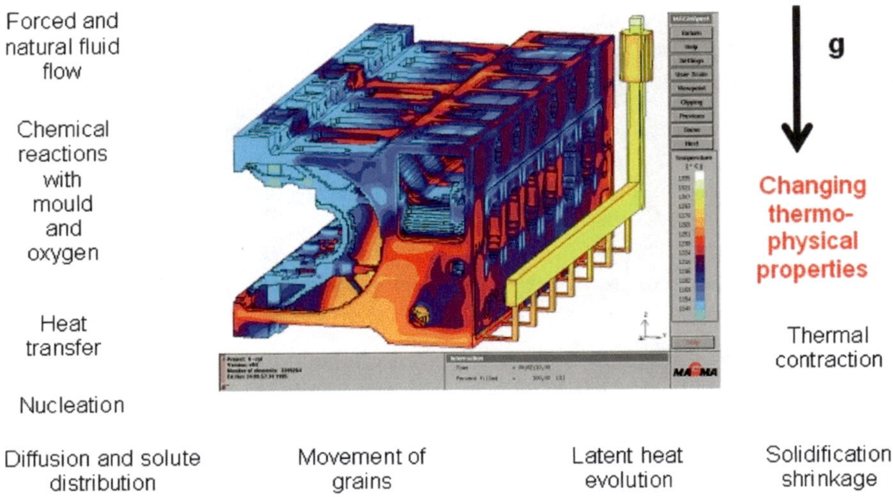

Fig. 3.10 A wide range of fundamental events during casting of complex components, here a car engine with varying local temperatures

On this basis, the quantitative numerical simulation of casting and solidification processes is increasingly demanded by manufacturers, compared to the well-established but time-consuming and costly trial-and-error procedure. It provides a rapid tool for the microstructural optimisation of high quality castings, in particular where process reliability and high geometric shape accuracy are important (see for example Fig. 3.10, exhibiting cast structural components and the temperature distribution during casting of a car engine block). Any improvement of numerical simulation results in an improved control of fluid flow and cooling conditions that enables further optimisation of the defect and grain structure as well as mechanical stress distribution. Through the control of unwanted crystallization events it even becomes possible to produce completely new materials with a controlled amorphous (glassy) or nano-composite structure.

Areas of research in which studies on critical pending questions have been defined, are described below.

Crystal Nucleation and Growth

During processing from the melt, for example for casting, welding, single crystal growth and directional solidification, crystal nucleation and growth is in most situations the first step achieved by cooling a liquid below its thermodynamic equilibrium solidification (liquidus) temperature to form crystalline nuclei of nanometer dimensions that subsequently start to grow. Alternatively, when the formation of nuclei fails, or the growth of nuclei is very sluggish, there is formation of a metallic glass at the glass transition temperature.

Indeed, the basic understanding of the fundamentals underlying the nucleation of crystals from the melt is in general limited to pure substances under well-controlled conditions. As an example, the heterogeneous nucleation rate $I(T)$ is typically described by [131].

$$I(T) = \frac{A}{\eta(T)} \exp\left\{ -\frac{16\pi\,\sigma^3}{3\,k_B T\,\Delta G_V^{lx}(T)^2} \right\}$$

with $\eta(T)$ the viscosity, σ the interfacial tension between the liquid and solid phase being formed at a heterogeneous nucleation site and $\Delta G_V{}^{lx}$ the difference of the Gibbs free energy between the undercooled liquid and the nucleating crystalline phase as a function of temperature with $T \leq T_{liq}$. For most alloys of actual interest these thermophysical properties are basically unknown.

For example, Fig. 3.11 exhibits experimental data of the shear viscosity η of liquid iron as a function of temperature exhibiting a large experimental scatter. The data for pure iron – the basis for any steel production – vary by about 50% [123]. Whereas a relatively large set of data exists for low temperature melts, the data for

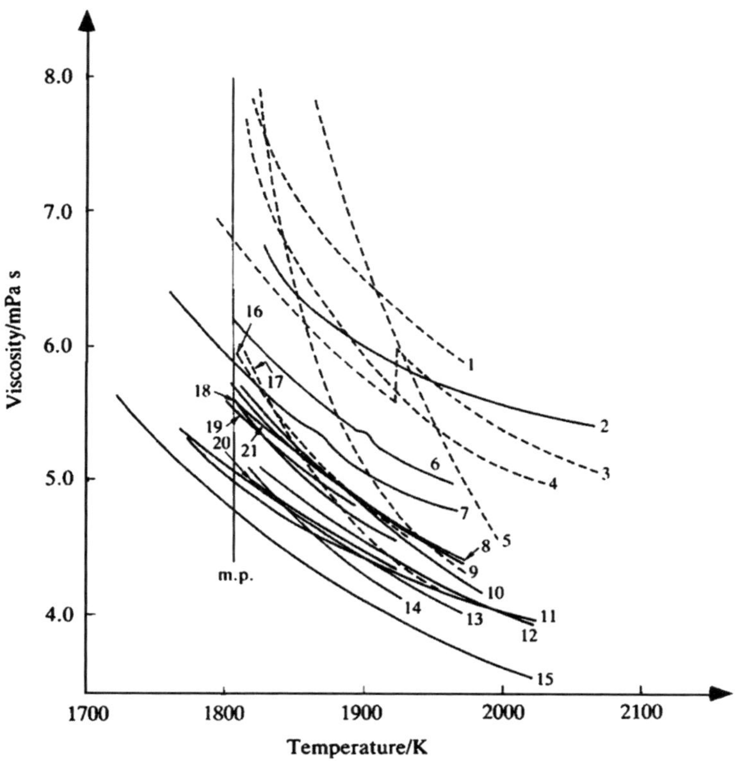

Fig. 3.11 Viscosity of 'pure iron' as a function of temperature from a number of different sources exhibiting a difference of about 50% near the melting point (Source: Iada and Guthrie [123])

liquids at high temperatures are rather scarce for pure metals and basically non-existent for complex alloys.

It follows that classical nucleation theory is based on many, sometimes unrealistic assumptions, failing to adequately include coupling between the nucleation barrier and structure of the initial phase and, thus, generally failing to describe complex multi-component alloys, which are used for engineering components.

(Bulk) Metallic Glass Formation

If the crystal nucleation rate is sufficiently low and if the growth of nuclei is sufficiently slow over the entire range of the undercooled liquid below the liquidus temperature, the liquid eventually freezes to a non-crystalline solid – a glass.

Glasses have been manufactured from silicon and related oxides for thousands of years. More recently, by developing new alloys and processing techniques it has become possible to produce more and more materials in an amorphous form which have superior properties in comparison with their (poly-)crystalline counterparts including materials with covalent (Si-based) or van der Waals (polymer) bonds.

In particular, new metallic glasses that can now be produced in large dimensions and quantities – so called bulk metallic glass (BMG) or supermetals – are emerging as an important industrial and commercial material, superior to conventional Ti-, Al- or Fe-based alloys. They are characterized by several times the mechanical strength (about 2 GPa) in comparison with conventional materials, excellent wear properties and corrosion resistance due to the lack of grain boundaries [71] (see Fig. 3.12).

These advanced materials can now be produced with dimensions of several ccm by casting or zone melting techniques. For a number of pseudo-eutectic Ti-Zr-based alloys cooling rates as low as 1 K/sec. are sufficient to obtain high levels of undercooling and produce a bulk glass with tailored properties, e.g., low weight and high strength. Most recent advances can be found in electronic casings for the

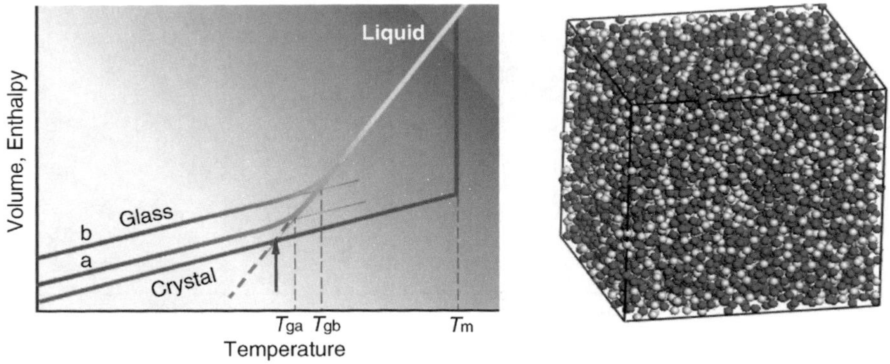

Fig. 3.12 Volume and enthalpy of a glass-forming alloy as a function of temperature and undercooling (*left*) and atomic structure in an MD simulation of a Zr-Cu glass [124] (*right*)

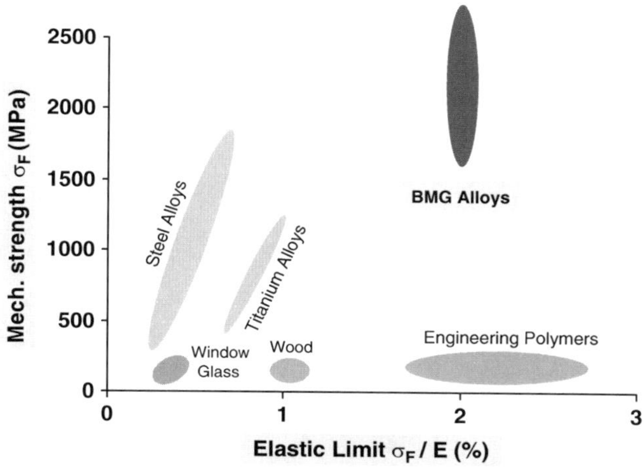

Fig. 3.13 Mechanical properties of BMG in comparison with more traditional materials, such as steels and Ti alloys, window glass, wood and engineering polymers (Source: Liquidmetal Technologies)

computer and communication industries, biomedical devices and implants as well as micromechanical parts.

A further interesting aspect is the controlled patterning of the surface by re-melting of the glass and patterning the surface in the temperature range of the undercooled liquid (exhibiting very small viscosities). Surface engineered topographies have been widely explored for silicon and transformed the electronics industry. Polymers can also be patterned and utilized, e.g. for optical applications, to control cell interactions, and drag reduction. The patterning of metals has been very limited due to their high flow stresses in their solid state and their grain size limitation. BMGs lack an intrinsic length scale and – unique among metals – can be thermoplastically formed (TPF) very similar to plastics. It has been demonstrated that this TPF processing opportunity can be utilized for the patterning of BMG surfaces as shown in Fig. 3.13 [138].

Intricate surface patterns have been applied covering features from 10 nm up to millimetres. Technologically, these patterns are most exciting since they can be optimized to affect wetting, adhesion, (foreign body) cellular response, and a vast range of optical properties (Fig. 3.14).

Columnar Growth and Columnar-Equiaxed Transition (CET)

Directionally solidified superalloy turbine blades, that are commercial cast-to-shape products, are columnar dendritic single crystals [83]. The mechanical performances of these blades depend on the fineness and regularity of the dendrite array. The blades can deteriorate rapidly if dendrite misalignment, parasitic

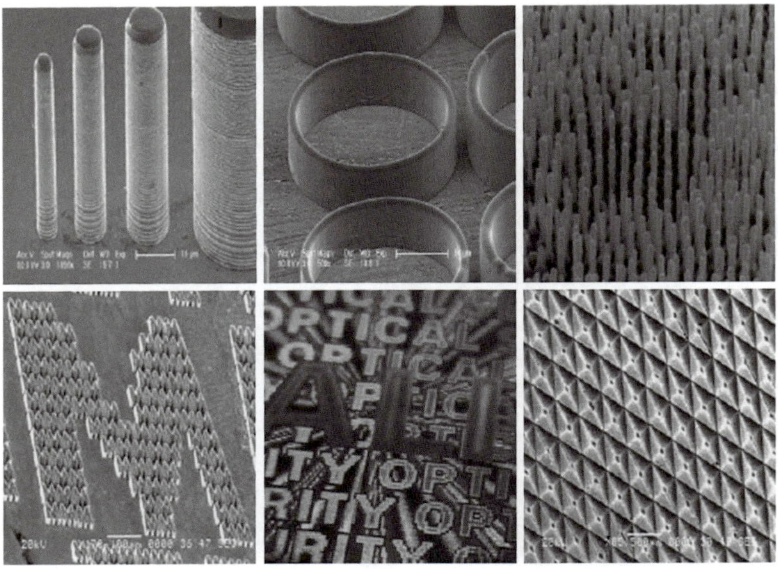

Fig. 3.14 Free standing micro and nanopillars and intricate surface patterns produced by a new technique down to the nanometer scale in BMG surfaces (Source: Yale University, USA)

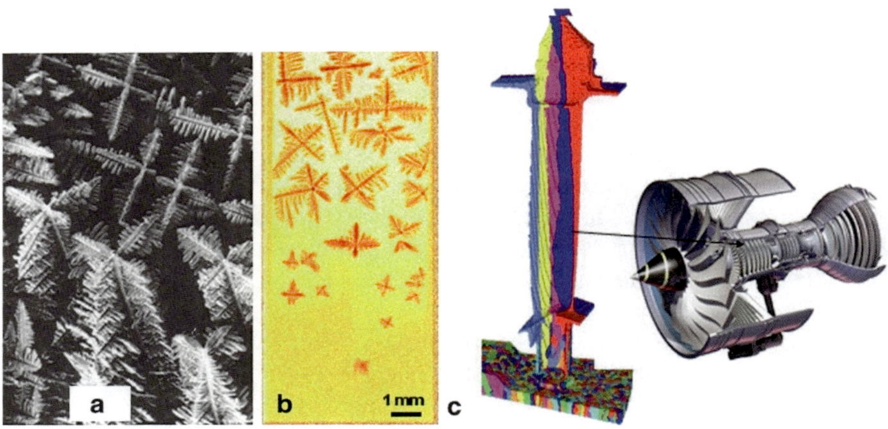

Fig. 3.15 (**a**) Columnar dendritic growth in a directionally solidified Co-Sm-Cu peritectic alloy showing primary and secondary arms; (**b**) Equiaxed grains growing in the melt during isothermal cooling down of an Al-4 wt% Cu alloy observed in situ by radiography at ESRF; (**c**) THERCAST® simulated 3D grain-structure in a turbine blade geometry produced by investment casting

nucleation of new grains or freckling induced by severe solute-driven convection occur during casting [73].

An example is shown in Fig. 3.15c with a simulated 3D grain-structure in a turbine blade geometry in comparison with dendritic structures in Co- and Al-based alloys (see Fig. 3.15a, b) [84]. This example shows that a better knowledge of equiaxed growth is imperative for the further improvement of a cast turbine blade. For years now, studies have thus been carried out at the minute level of the free growth of a single dendrite, at the intermediate level of the envelope of a single grain embedded in its neighbourhood, and in parallel at the collective level of a set of interacting grains. Any fluid flow inference makes every type of study much more convoluted. Also, it will be of value to thoroughly investigate and precisely determine the conditions of the transition from columnar dendritic growth to equiaxed crystal formation avoiding grain sedimentation [61]. A full and thorough analysis of this problem however requires an accurate knowledge of the relevant thermophysical properties of the melt.

3.1.4.3 Specifics of Low-Gravity Platforms and Facilities for Materials Science

Fresh insight into the stable and metastable undercooled liquid state and metallic alloy solidification can be gained with the potential of engineering novel microstructures. To perform these experiments it is important to have access to extended periods of reduced gravity (also see [90] GoSpace).

Parabolic Flights

Parabolic flights generally provide about 20 s of reduced gravity. For materials science experiments at high temperatures, this time is barely sufficient for melting, heating into the stable liquid, and cooling to solidification of most metallic alloys of interest in a temperature range between 1,000–2,000 C. Surface oscillations can be excited by a pulse of the heating field, and the surface tension and viscosity are obtained from the oscillation frequency and damping time constant of the oscillations, respectively. Processing must be performed in a gas atmosphere under convective cooling conditions. Under these conditions, however, it is not possible to reach thermal equilibrium of the melt.

TEXUS Sounding Rocket Processing

TEXUS sounding rockets offer a total of 320 s of reduced gravity which is typically split between two different experiments. As compared to parabolic flights, the microgravity quality is improved by far. Stable positioning and processing of metallic specimens with widely different electrical resistivity and density was achieved in different sounding rocket flights with an adapted electromagnetic levitation device. In Fig. 3.16, the temperature-time profile of the Fe alloy processed successfully in TEXUS 46 EML-III is shown as an example. However,

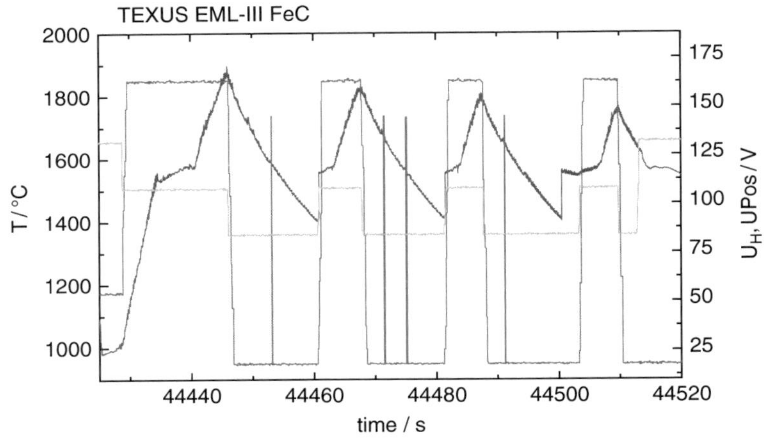

Fig. 3.16 Temperature-time profile of the FeC alloys processed on the TEXUS 46 EML-III sounding rocked flight with Temperature scale (*left*) and heater and positionner voltage (*right ordinate*)

also under these conditions thermal equilibrium or steady state conditions of the liquid phase cannot be attained due to the limitations in processing time.

Long-Duration Microgravity Experiments on ISS

A series of microgravity research projects is now commencing onboard the ISS in a number of multi-user facilities afforded by major space agencies, such as the European Space Agency (ESA) in the Columbus module. The short microgravity time on parabolic flights and in TEXUS sounding rocket experiments necessitates forced convective cooling to solidify the specimen before the end of the microgravity time. As a consequence, long-duration microgravity measurements are needed for accurate calorimetric and thermal transport property measurements over a large temperature range. It can be expected that the following properties necessary for a full and thorough analysis of solidification processes will be performed at varying temperatures (ThermoLab-ISS programme: an international effort sponsored by several national agencies ESA, DLR, NASA, JAXA, CSA and others):

- viscosity – using the oscillating drop technique
- surface tension and density – same as above
- total hemispherical emissivity – using modulation calorimetry and measurement of external relaxation times
- specific heat – same as above
- other calorimetric properties
- electrical conductivity – using inductive methods to measure the impedance of a pick-up coil surrounding the sample

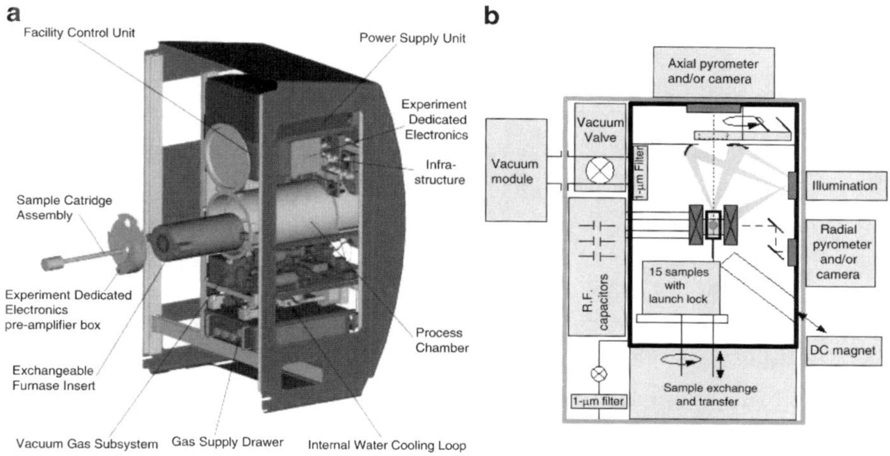

Fig. 3.17 Schematic presentation of (**a**) the Materials Science Laboratory and (**b**) the electromagnetic levitator reaching temperatures up to 2,200°C

- thermal conductivity – using the Wiedemann-Franz law
- alloy melting range and fraction solid – using modulation calorimetry

Furthermore, in Materials Science both directional and isothermal solidification experiments will be performed in the Materials Science Laboratory (MSL, see Fig. 3.17a) using dedicated furnace inserts, with the possibility of applying external stimuli such as a rotating magnetic field to force fluid flow in a controlled manner. The electromagnetic levitator (EML) will enable containerless melting and solidification of alloys and semiconductor samples. The EML is equipped with highly advanced diagnostic tools as shown schematically in Fig. 3.17b.

3.1.4.4 Experimental Performance and Perspectives

The lack of information for commercial materials as well as materials of fundamental interest is a result of the experimental difficulties generally arising at high temperatures, i.e., the dominance of entropic effects. By eliminating the contact between the melt and a crucible, accurate surface nucleation control and the synthesis of materials free of surface contamination become possible. For highly reactive metallic melts, electromagnetic levitation is a well-developed containerless technique which offers several advantages over alternative levitation methods (electrostatic levitation, gas-phase levitation) due to the direct coupling of the high-intensity radiofrequency electromagnetic field with the sample.

As a demonstration, Fig. 3.18 shows a comparison between a specimen levitated in ground-based em-levitation (left) and a liquid specimen positioned under reduced gravity conditions in an em-levitation device on board a parabolic flight

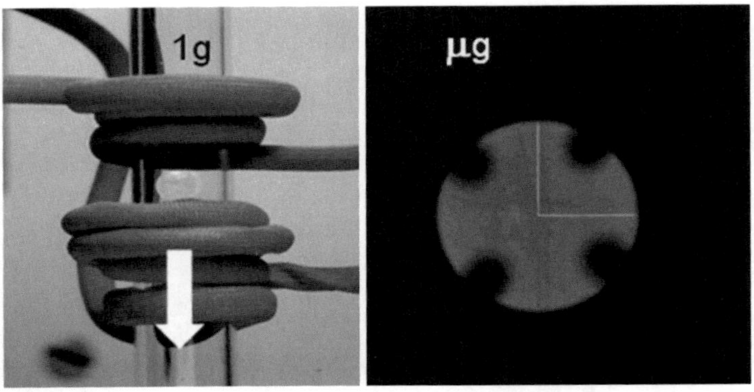

Fig. 3.18 EML processing on the ground (*left*) and in reduced gravity on board a parabolic flight (*right*) allowing investigations of fully spherical samples with a range of analytical equipment

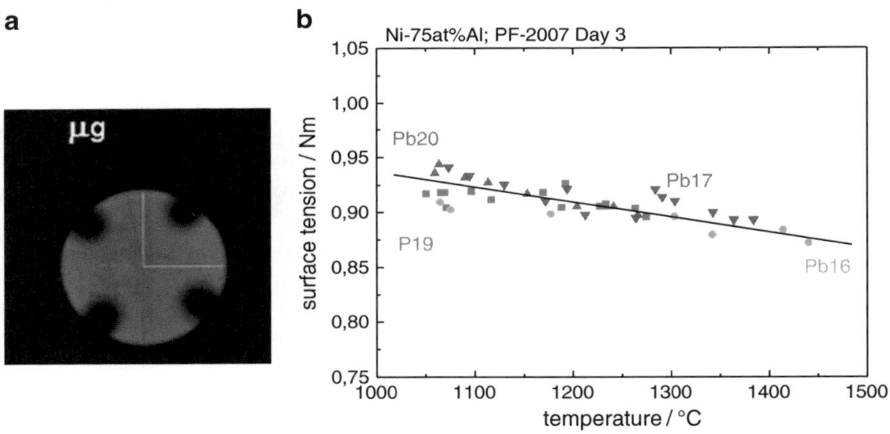

Fig. 3.19 (**a**) Video image of a fully spherical liquid sample of a NiAl alloy in EML obtained in a parabolic flight for surface tension and viscosity measurements of liquid metallic alloys, (**b**) Surface tension of a drop of molten Ni-75 at% Al between 1,050°C and 1,450°C

(right). As compared to the specimen levitated on ground, the specimen positioned under reduced gravity exhibits no detectable deviation from a spherical shape.

Using the Electromagnetic Levitator, surface oscillations of the liquid hot drop with diameter of 8 mm (Fig. 3.19a) can be, for instance, introduced by an electromagnetic pulse and the results analyzed by a high-speed high-resolution video camera. Among the results, Fig. 3.19b shows the surface tension as a function of temperature in the range 1,050–1,450°C of liquid Ni - 75 at % Al processed under low gravity on four parabolic flights with 10 s of processing time each.

3.1.4.5 Materials Alloy Selection

For a wide range of new products in the industrial production chain, solidification processing of metallic alloys from the melt is a step of uttermost importance. On this basis, several types of alloys have been selected for the following fields of usage. Examples are:

- Turbine blades for land-based power plants and for jet engines sustaining high temperatures and high stress levels
- Low-emission energy-effective engines for cars and aerospace
- Functional materials with improved performance (electronic transport, soft/hard magnetic magnets, catalytic efficiency, hydrogen storage, thermoelectric power)
- Supermetals (bulk metallic glass) such as thin sheets for electronic components with ultimate strength to weight ratio
- Biocompatible medical implants such as hip and tooth replacements
- New low-weight and high-strength materials for space exploration and future space vehicles.

As such, these alloys contain several different metallic and (sometimes) non-metallic elements, such as Ti, Ni, Fe, Cu, Al etc. with B, C, N additions. In particular the following alloys will be of major interest:

Ti-alloys such as Ti6Al4V (the so called Ti64 base alloy) and Ti-aluminides are of great interest to the aerospace industry (see Fig. 3.20). This group of alloys is constantly being further developed by the addition of refractory elements such as tungsten, rhenium and zirconium which pushes the liquidus temperature up to 1,700°C and more. In this temperature range no quantitative thermophysical property measurements by conventional methods are possible.

Ni-based superalloys. Ni-based superalloys have long been the workhorse for power generation in land-based turbines and for jet engine propulsion. Their γ/γ' microstructure is characterized by cubic precipitates of an ordered cubic phase, γ', in a disordered matrix with a cubic crystal structure of almost identical lattice

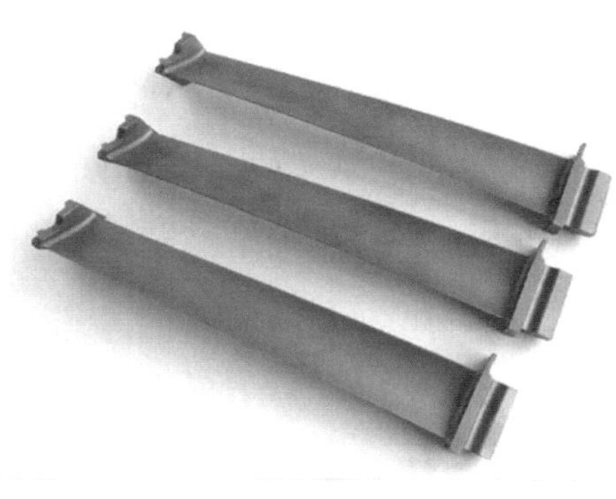

Fig. 3.20 Large cast Ti-Al-Ta(Nb) alloys for advanced turbine blade technology [96]

constants, the γ-phase, which makes these materials capable of functioning at the highest temperature of any materials in routine service. This basic structure is constantly being further developed, with so-called third and fourth generation Ni-based alloys.

Fe-based alloys. These steels are used in land-based turbines for energy production, in jet engines, the automotive industry, advanced fusion concepts and more. They are also very interesting for the study of competing nucleating kinetics of different phases. Steel casting alloys show complex solidification steps with primary metastable ferritic phase formation, rapidly followed by a transformation to the stable austenite. This transformation has been shown to be significantly influenced by liquid convection. Maintaining microstructural control requires development of casting models which include these effects.

Cu-based alloys. These alloys are designed to combine excellent electrical conductivity with good mechanical properties, such as sheet material or wires for electronic applications. The key problem to be solved is porosity after casting resulting from two sources: (i) shrinkage porosity is caused by the change in volume (or density) when a liquid solidifies, and (ii) gas porosity is caused by the marked decrease in the solubility of gases when an alloy solidifies and gas bubbles are formed when the pressure exceeds one atmosphere.

Bulk metallic glasses. Bulk metallic glasses represent a new development in materials science. These materials possess superior mechanical properties compared to crystalline conventional materials. The first generation of these materials was based on combinations of early and late transition metals with further addition of, for example, Al and B. New, more lightweight compositions have been developed in recent years. For the improvement of bulk metallic glass formation a thorough understanding of the kinetic and thermodynamic properties affecting nucleation and phase formation is required, as well as refined models of crystal nucleation from deep undercooling.

New Materials, Products and Processes

Apart from the knowledge of thermophysical properties discussed above, a full analysis of solidification and crystal growth phenomena requires an understanding of fundamental mechanisms. Using microgravity, considerable progress has been achieved in this respect through the interplay between benchmark experiments in μg and 3D computer modelling. Prominent examples are the prediction and observation of Columnar-Equiaxed Transition (CET) and equiaxed single grain growth from an undercooled alloy. There is a general trend now to move away from simple systems to multi-component, multiphase alloys, including stable and metastable monotectics, leading eventually to *in-situ* functional materials.

A major achievement was also the development of *in-situ*, real-time diagnostic tools, such as time-resolved X-ray radiography. Among many results the three most significant achievements in the field can be summarized as follows:

- fractal-like agglomeration of Ni-nanoparticles for fuel cell applications [97],
- single grain growth in levitated undercooled drop (Fe-C-Mn)/experimental observation and 3D modelling,
- advanced intermetallics (TiAl, NiAl) and quantitative prediction of structures (CET, dendrite morphology).

Several series of ground and TEXUS/MAXUS sounding-rocket experiments were carried out within the MICAST and CETSOL, MAPs and IMPRESS projects to pave the way to the utilization of the Low Gradient Facility and Solidification and Quenching Facility in MSL. Unexpectedly, it was found that fluid flow, either driven by gravity or forced by a magnetic field, was enhancing dendrite arm coarsening, although tip radius (Fig. 3.21b) was not affected. Benchmarking of the columnar-to-equiaxed transition has begun on Al - 7 wt% Si alloys with MAXUS7, which are suggesting that the presence/absence of CET should be linked to the effectiveness of melt inoculation by dendrite-arm fragmentation. For the sake of comparison, development of phase-field modelling is pursued to reach quantitative numerical prediction from the scale of the dendrite tip (Fig. 3.21a) to the scale of cooperative array growth (Fig. 3.21c).

The IMPRESS Integrated Project comprised about 40 industrial and academic research groups from 15 countries in Europe as shown in Fig. 3.22a. The project aimed to elucidate the relationship between the processing, structure and final properties of new intermetallic alloys, such as TiAl and NiAl. Indeed, these intermetallic alloys, due to their crystal structures and strong chemical bonds, have many attractive mechanical, physical and chemical properties that make them of value for various applications. For instance, a new castable intermetallic Al-Ti alloy was developed for investment casting of lightweight high-strength turbine blades with 40 cm length for next-generation aero engines (Fig. 3.22b), which will contribute to reducing fuel consumption. Also, first fractal-like aggregates of catalytic Ni (Fig. 3.22c), of potential relevance for high-performance fuel cell applications, were produced in low-gravity where collapse under own-weight is suppressed.

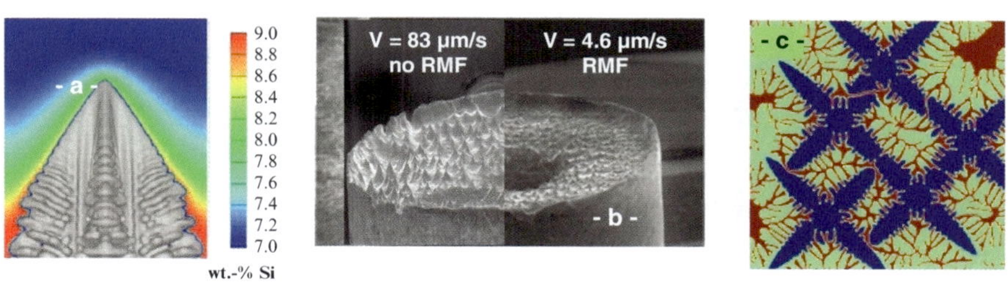

Fig. 3.21 (**a**) 3D-Phase-field simulation of AlSi7 dendrite with Si solute field (ACCESS) (**b**) Decanted solid – liquid interfaces of Al-7Si Mg samples directionally solidified without/with Rotating Magnetic Field (ACCESS), revealing the dendrite tips (**c**) 2-D phase-field simulation of Ni-Al solidification showing primary Ni2Al3 dendrites (*blue*), NiAl3 peritectic (*green*) and remaining liquid (*red*/*brown*) (Source: A. Mullis, Univ. Leeds)

Fig. 3.22 IMPRESS Integrated Project. (**a**) Geographical distribution of IMPRESS partners in Europe (**b**) Light-weight high-strength TiAl turbine blades (**c**) Fractal-like aggregate of catalytic Ni

3.1.4.6 Summary and Outlook

Materials Science is still an emerging field relevant for society and industrial development. Among several different disciplines, liquid processing and control of resulting structures and products has been defined as a key issue with a huge turnaround. Further improvement of industrial design for our society, in order to reduce energy consumption and develop green technologies, requires new materials and processes where science in space is a key element. The following materials and process developments have been defined and can be summarized as follows:

Advanced Materials

- Atomized and spray-formed powder (structural, catalysis)
- Bulk metallic glasses/foams (Ca-, Mg-, Ti-based; monolithic, foam, nanowires, surface patterns)
- Complex metallic alloys (intermetallics, multicomponent alloys, nanostructures, high entropy, etc.) and high temperature alloys
- Control of grain structure in solidification processing coupled to quantitative predictive 3D modelling
- Energetic pyrotechnic metals (nano-Al, Mg, Fe)
- Fuel cell catalysts (Ni, PGMs)
- Heat exchangers/Heat sinks (Al, Cu, Na, nano-Au)
- Hollow sphere metal foams (Ti, Al)
- Metal matrix composites (containing nanofibres, CNTs)
- Nano-porous de-alloyed metals (Various)
- Metal composites with inert nano-oxides (Mg, Al)
- Shape memory alloys (Ni-Ti)

Advanced Processes

- Joining of interfaces (electron beam and laser welding)
- Power ultrasonic processing
- Power ultrasound processing of liquid metals
- Structural modifications by deep undercooling
- X-ray imaging for liquid processing

Acknowledgments The financial support from ESA and DLR and the scientific cooperation within the International ThermoLab Team (MAPs and Topical Teams), GoSpace and with B. Billia (Marseille, F) are gratefully acknowledged.

3.1.5 Space Weather by Hanna Rothkaehl

Hanna Rothkaehl

Space Research Center Polish Academy of Sciences, Bartycka 18A,
00-716 Warsaw, Poland

3.1.5.1 Introduction

Our highly technological everyday life is exposed to the effects of space weather. The definition of space weather elaborated by COST 724 Action reads as follows: *Space weather is the physical and phenomenological state of natural space environments. The associated discipline aims, through observation, monitoring, analysis and modelling, at understanding and predicting the state of the Sun, the interplanetary and planetary environments, and the solar and non-solar driven perturbations that affect them, and also at forecasting and nowcasting the potential impacts on biological and technological systems.* Each solar storm, for example, generates intensive showers of particles and gigantic currents in the ionosphere which induce major alterations in the geomagnetic field. Electric conductors in the changing magnetic field, whether cables, pipes or seawater, run currents called geomagnetically-induced currents, GICs.

3.1.5.2 UV Solar Radiation

The sun radiates in a wide range of electromagnetic wavebands. The ultraviolet (UV) electromagnetic emissions are located between visible and X-ray frequency bands within wavelengths from 10 to 400 nm and energies from 3 to 124 eV. The solar atmosphere emits UV radiation which varies strongly during the solar cycle and, in consequence, drives variations in the near-Earth environment. Ultraviolet

Radiation from the sun is an area of interest to both the scientific community and the general public. UV radiation interacts strongly with the ionosphere and atmosphere layers. Moreover, UV flux is a major driver of stratospheric chemistry and can affect increases in ionosphere-influenced temperature. On the other hand, man-made pollution can cause changes in the actinic flux of ultraviolet UV radiation. Observations and numerical models show that UV-scattering particles in the boundary layer accelerate photochemical reactions and smog production, but UV-absorbing aerosols such as mineral dust and soot inhibit smog production. The results could have major implications for the Earth's climate. The Earth's atmosphere blocks most UV radiation (97%); UV-C (100–280 nm) is screened out by stratospheric ozone and oxygen at around the 35 km altitude and UV-A (280–315 nm) radiation represents only one percent of the radiation reaching the Earth's surface, but UV-B (315–400 nm) radiation is largely responsible for sunburn and skin cancer and UV-Bs are hazardous to the eyes. On the other hand, UV-B exposure induces the production of vitamin D which is necessary for life. UV radiation can also cause degradation of solar cells, some EPROM (erasable programmable read-only memory) modules, polymers, pigments and dyes used for space material and technology purposes. An increase in UV radiation makes the atmosphere expand, moving satellites to lower orbits and slowing them down, even causing them to fall to Earth. The crash of the space station Skylab in 1979 was a famous example of this.

Since the first satellites were located in orbit, UV radiation has been monitored for the Sun, the near Earth environment and the universe. The relevant solar UV studies began in the late 1960s with sounding rockets and stratospheric balloon flights and have had a space segment since SPACELAB-1 in 1983. Recently, solar UV images have been obtained from instruments located on board the SOHO and STEREO spacecrafts. In the near future, ESA will launch a Proba-2 satellite to demonstrate innovative technologies as part of the In-orbit Technology Demonstration Programme. Aboard will be LYRA instruments for monitoring four bands across a very wide ultraviolet spectrum.

The UV diagnostics located on the ISS are crucial for mitigating the hazards to astronauts and space technology. The SOLAR package and its three instruments (aboard COLUMBUS since February 2008) can study solar irradiation with unprecedented accuracy for a series of Sun observation cycles. The data are stored at the Belgian Institute for Space Aeronomy.

In order to provide a representative service of UV radiation, great efforts are being made to construct a European network for the measurement of visible UV radiation on the ground as well.

Recently, a ground-based observation system provided by SUVIM (the Belgian Institute for Space Aeronomy) was established (Fig. 3.23). The systematic archiving of data began only in 1993 and now includes five ground stations: Uccle (Brussels), Mons, Redu, Mol and Ostend. All five stations are located in Belgium and a sixth station in Diekirch (Luxembourg) will soon be included in the project. The instruments, stations, current applications and evolution are described at http://www.aeronomie.be/uv/globaluv/index.php. The data are already being used in an

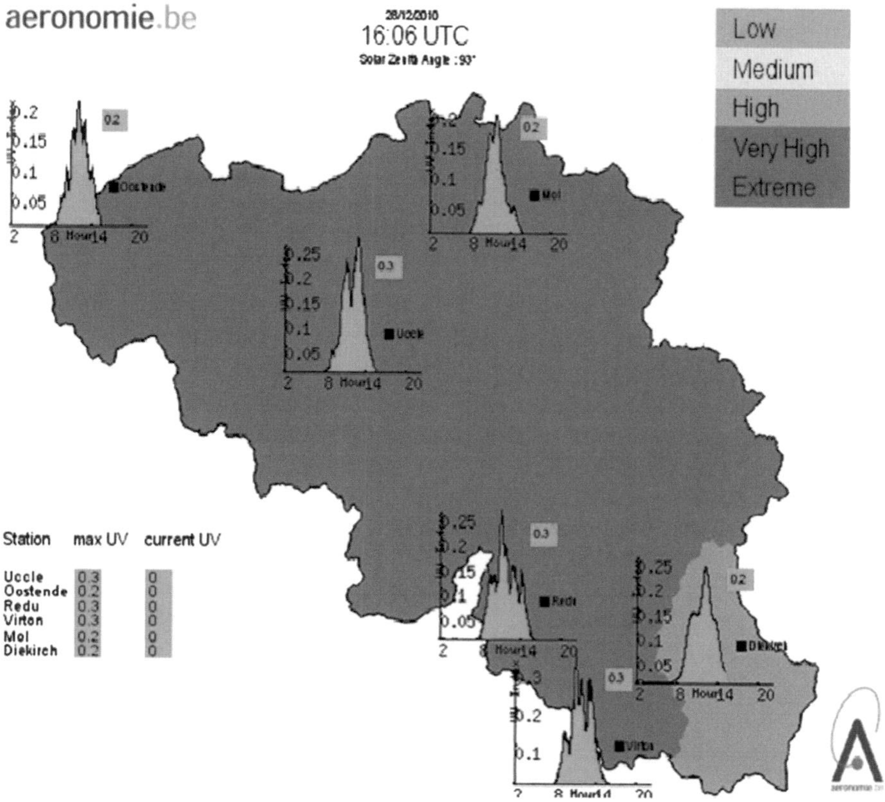

Fig. 3.23 Example of SUVIM service, UV index for 28 December 2010

operational service distributing the UV indexes and solar irradiances in real time together with ancillary meteorological parameters.

3.1.5.3 Electromagnetic Earth Environment

The Sun continuously emits two kinds of radiation: electromagnetic and corpuscular. The electromagnetic emissions radiated by the Sun can be one of the major drivers of change in the Earth's environment. A description of near-Earth space plasma behaviour is a current subject of investigation, both as a constituent of the geophysical environment and as an element of physical processes in which particles and waves participate. Whistler-mode hiss, chorus, and EMIC waves are now believed to be involved in the processes of particle acceleration and loss at rates greater than hitherto estimated. There is considerable interaction between the electromagnetic waves emitted by the Sun and the Earth's atmosphere, although it is not yet fully understood.

Recent studies have also shown that human activity can perturb the near-Earth space environment. They indicate that the observed significant increases in electromagnetic turbulence over Europe and Asia are caused by the permanent pumping of electromagnetic waves into the ionosphere by a system of broadcasting stations. This effect can be intensified by precipitation of energy particles from the radiation belts.

The correlation between earthquakes and anomalous bursts of trapped particles, precipitating from the lower boundary of the inner radiation belt was observed as an intensification of HF wave activity by *in-situ* topside experiments. It seems that changes in the magnetic flux tube topology, correlated with seismic activity, can lead to an increase in the precipitation of energetic electron fluxes and, as a consequence, lead to an excitation of the HF whistler mode.

Theoretical investigations and some space-borne and ground-based experiments have shown that natural whistler waves generated by a lightning discharge in the Earth's atmosphere can accelerate the trapped radiation belt electrons.

In order to mitigate the radiation hazards to space-borne experiments and space exploration programs, we need to increase the number of measurement points in space and on the Earth's surface and construct an improved model to describe different types of physical processes driven by fluxes of energy particles and wave-particle interactions with space plasma.

This should help us to achieve a comprehensive understanding of the combined natural and anthropogenic plasma processes and their interactions with geospace. The community of users potentially benefiting from these investigations include the civilian and the defence sectors, the aviation industry, the satellite industry, HF equipment manufacturers, HF broadcast and communication services and trans-ionospheric radio operations (GPS, GLONASS, Galileo).

Conducting electromagnetic monitoring aboard the ISS (International Space Station) requires both the elaboration of observation methodology and the design of the corresponding experimental equipment. To achieve this goal, the "OBSTANOVKA-1" experiment (in English, "ENVIRONMENT-1"), located on the ISS, will be carried out to provide a databank of electromagnetic fields and plasma-wave processes in near-Earth space. The Plasma-Wave Complex (PWC) instrument has been designed and constructed in the framework of international cooperation.

The PWC contains a combined wave sensor, flux gate magnetometer, Langmuir probe, spacecraft potential monitor, plasma discharge stimulator, Correlating Electron Spectrograph (CORES), Radio Frequency Analyzer (RFA), signal analyzer and sampler, data acquisition and control unit, telemetry information storage block and ground support equipment. The PWC instrument will be located in the Russian Segment of the ISS in spring 2011. This complex set of instruments will provide a unique chance to (Fig. 3.24):

- determine the spectral density of electromagnetic, electrostatic and magnetic field fluctuations in a range of frequencies from fractions of hertz up to tens of megahertz resulting from various natural and artificial influences;

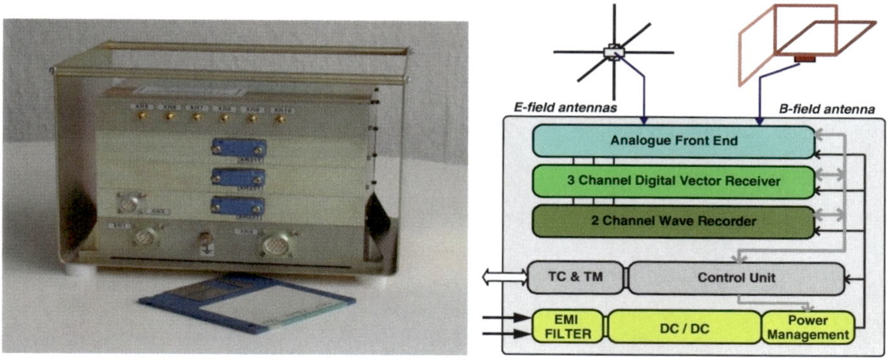

Fig. 3.24 Photograph of RFA analyser and block diagram

- measure the vectors of intensity of magnetic fields and field-aligned currents (FACs);
- determine spectral fluctuation of the charged particle flows and density;
- estimate the conformity of measured electromagnetic fields to the operational requirements of space engineering products and technology, service systems and useful payload, and estimate electromagnetic changes during the docking process.

The new digital Radio Frequency Analyzer, which forms part of the PWC, was designed to monitor and investigate electromagnetic radiation from 100 kHz up to 15 MHz. This joint Polish (SRC PAS, Warsaw)–Swedish (SISP, Uppsala) instrument was designed to diagnose electric and magnetic vector fields, **E** and **B**, respectively, in the space and time domain. The level of electromagnetic noise detected at ionospheric altitudes depends on the geophysical condition, properties of the different Earth regions, influences from the Earth's surface and the interaction between the body of the spacecraft and sounding plasma.

The registered data will be available via the ULISSE service portal. http://ulisse.cbk.waw.pl/

3.1.5.4 High Energy Radiation

Galactic cosmic rays (GCR) and solar energetic particles (SEP) cause that, during human flight in space and on polar routes, flight personnel and passengers are consequently exposed to higher radiation intensity and have to be monitored for compliance with safety rules. The astronauts from the ISS are exposed to a wide spectrum of radiation particle and energies. The mapping of radiation levels throughout the internal environment of the International Space Station and in the immediate vicinity of each crewmember is one of the major experimental issues of spaceflights.

One example of such a monitoring instrument is the LIULIN-type Deposited Energy Spectrometer DES, designed and contracted by the Bulgarian Academy

Table 3.1 Examples of dosimeter experiments

Instrument	Operating place	Time frame	Owner/PI
Radiometry Dosimetry System LIULIN		1988–1994	Solar-Terrestrial In-fluences Laboratory, Bulgarian Academy of Sciences (STIL-BAS) - Collaboration with IMBP-Russia
LIULIN-E094	Human Research Facility of Expedition Two Mission 5A	May, 2001 – 21/July, 2001	Solar-Terrestrial In-fluences Laboratory, Bulgarian Academy of Sciences (STIL-BAS) - Collaboration with DLR
LIULIN-MKC	In the service radiation monitoring system of the Russian segment of ISS	Launched in September 2005 and operating since August 2008	Solar-Terrestrial In-fluences Laboratory, Bulgarian Academy of Sciences (STIL-BAS) - Collaboration with IMBP-Russia and NPO-Energia

of Sciences STIL BAS, which measures the spectrum (in 256 channels) of the deposited energy in a silicon detector from primary and secondary particles at aircraft altitudes, Low Earth Orbits (LEO), outside the Earth's magnetosphere and on the surface of the planets of the Solar system. Table 3.1 [276] lists examples of dosimetric experiments performed by LIULIN-type detectors. Moreover, the Radiation Risk Radiometer-Dosimeter (R3D) is designed to measure the radiation climate during the SPORES long-term life science experiment on the ISS. R3D is a low-mass, low-dimension automatic device which measures solar radiation in four channels and cosmic ionizing radiation in 256 channels. Other instruments onboard the ISS are the DOSTEL (DOSimetricTELescope) instruments, which were operated inside BIORACK on the Space Shuttle inside the ISS during Increment 2 and outside the ISS (Advanced DOSTEL) as part of the EuTEF platform. Two Dostel instruments performed measurements inside the ISS from July 2009 to June 2011 and will continue these measurements as part of the DOSTEL 3D experiment from March 2012.

The aim of the MATROSHKA project was to study the depth dose distribution of the different components of the orbital radiation field on different sides of the organs of astronauts during an extravehicular activity (EVA). The MATROSHKA experiments consist of a simulated human body (head and torso), called the Phantom, equipped with several active and passive radiation dosimeters. There were two series of the MATROSHKA Project between 2004 and 2009. The owners of the data are DLR (Germany), TUV (Austria) and IF PAN (Poland) and the available data are stored on the webpage http://www.fp7-hamlet.at.

The main task of the IMPULS experiment (in 2009, located aboard the Russian segment of the ISS) was to monitor modification of ionospheric plasma by pulse plasma beams. The experiment was carried out by IZMIRAN (Russia). The studies focus on registration and examination of electrical field parameters near the ISS surface during various flight conditions; the response of electrical field conditions to the injection of a pulsed plasma beam near the ISS body; and discharge processes on the surface of the ISS. It is also planned to put the LIGHTNING-GAMMA experiment in the Russian segment in 2011 in order to monitor optical and gamma emissions related to thunderstorm activity. The experiment will contain three components of gamma and optical sensors with high temporal resolution.

3.2 Life Sciences

3.2.1 Astrobiology

Pascale Ehrenfreund[1], **Frances Westall**[2]

[1]Institute of Chemistry, Leiden University, Leiden, The Netherlands and Space Policy Institute, George Washington University, Washington, DC, USA
[2]Centre de Biophysique Moléculaire-CNRS-OSUC, rue Charles Sadron, 45071 Orléans cedex 2, France

3.2.1.1 Introduction

Astrobiology, the study of life in the universe, seeks answers to fundamental questions on the origin, evolution, distribution and future of life, wherever it may exist. Through interdisciplinary research that combines disciplines such as molecular biology, ecology, planetary science, astronomy, information science, space exploration technologies, and other fields, Astrobiology tries to find answers to the basic questions on how life began and evolved, if life exists elsewhere in the universe, and what the future of life on Earth and beyond may be?[1] For more than two decades astrobiology experiments have been performed successfully in Earth orbit and provided interesting new insights into the effects of the harsh space environment on biological and organic materials. The parameters that are most critical are increased radiation exposure (solar ultraviolet and ionizing radiation) and microgravity, neither of which can be accurately simulated in Earth laboratories.

3.2.1.2 Recent Astrobiology Platforms

The exposure facilities Long Duration Exposure Facility (LDEF), BIOPAN/FOTON, EXPOSE-E (EXPOSE-Eutef) and EXPOSE-R, as well as planned new

[1]http://astrobiology.arc.nasa.gov/roadmap/.

external payloads on the International Space Station (ISS), provide opportunities to study the evolution of biological and organic materials and chemical processes directly in Earth orbit, overcoming the operational difficulties of laboratory simulations [53, 65, 94, 118, 192, 211]. Investigations of large organic molecules, such as polycyclic aromatic hydrocarbons (PAHs) on BIOPAN 5 showed negligible destruction of those compounds during the two-week mission [65]. Experiments on the BIOPAN 6 mission measured the half-lives of several amino acids, urea and HCN polymers [94]. Space measurements differed compared to data derived from laboratory experiments and indicate the need for more data to reconcile these discrepancies.

EXPOSE-E, EXPOSE-R and planned new external payloads are multi-use space exposure facilities designed for astrobiology, radiation biology, and radiation dosimetry research in extreme environments, and material science investigations in space [59, 277], see Fig. 3.25. Extended space exposure allows us to collect data on multiple samples which can be extrapolated to other astrophysical environments.

The effect of microgravity and radiation exposure on biological material has been recently reviewed by Horneck et al. [119, 278]. Investigations on the germination of Arabidopsis seeds in space and simulated microgravity indicated that cell proliferation and cell growth are uncoupled in the space environment [164], where recent experiments have also shown that microbes can be more pathogenic. Increased virulence of Salmonella has been observed that may involve a global regulatory signal that affects gene expression [270].

The environment on the ISS allows for a realistic simulation of the alteration of organic compounds in the space environment. In the next decade ISS utilization

Fig. 3.25 The external exposure facility EXPOSE-R attached to the external platform URM-D was operational on the ISS until January 2011. Experiments on-board this facility investigate the effects of space radiation on organic and biological material (Source: ESA)

will be vital for advancing space research. The development of *in-situ* measurement technology provides an essential and timely element for the exploitation of the ISS.

NASA's Organism/Organic Exposure to Orbital Stresses (O/OREOS) nano-satellite was launched successfully to a high-inclination (72°), 650-km Earth orbit aboard a US Air Force Minotaur IV rocket from Kodiak, Alaska on November 19, 2010. O/OREOS measurements provide, for the first time, a real-time analysis of the photostability of organic compounds and biomarkers by taking weekly spectra using a UV spectrometer. Furthermore O/OREOS will obtain data on the survival and metabolic activity of two different microbes at three times during the 6-month mission using optical density measurements.

Experiments to test the effects of atmospheric entry on anologue Martian sedimentary meteorites using rocks embedded around the stagnation point in the heat shields of FOTON capsules have been conducted since 1999 [32, 34, 77]. Sediments are of interest because of their potential for containing traces of Martian life but no Martian sedimentary meteorites have yet been found. All flown experiments have shown the survivability of carbonates, sandstones, lacustrine muds and volcanic sediments, some of which contained organic biosignatures and fossil microorganisms. However, none of the sediments developed the characteristic black fusion crust typical of stony meteorites, although fusion did lead to the formation of a vitreous crust on the siliciclastic rocks (an igneous rock (dolerite) used as a control did exhibit a dark fusion crust), see Fig. 3.26.

The heat shock of atmospheric entry produced changes in the sediments similar to those experienced by stony meteorites, namely material loss due to melting, ablation, and decrepitation. Indeed, structural and compositional changes in certain minerals demonstrate that the minimum temperatures experienced were >1,783°C (the melting temperature of quartz). The STONE 1 and STONE 5 experiments consisted of flanged disks of rocks 1 cm thick. Live colonies of endolithic photosynthetic microorganisms from an Arctic desert environment were placed at the back of the STONE 5 experiment to determine whether such natural endoliths could

Fig. 3.26 Image of the FOTON capsule and one of the sediments exhibiting a white (not black) fusion crust after entry into the Earth's atmosphere (Source: ESA)

traverse space to inseminate another planet, as in the panspermia hypothesis. Since the 1 cm thickness was not sufficient to protect the organisms from the heat shock during atmospheric entry, in the STONE 6 experiment the rocks were milled into a dome shape that was 2 cm at the apex. Modelling based on mineralogical changes in the thickness of the sample show that 2 cm was still not enough and that a rock thickness of at least 5 cm is necessary to protect a potential living "hitchhiker" [77].

Cockell et al. [51] concluded that, since endolithic photosynthesizing organisms need to be close to the surface of a rock in order to have access to light, such organisms could not have survived atmospheric entry and therefore must have originated on Earth. Although the live organisms did not survive, fossil signatures of life in the sedimentary rocks (microfossils and biomarker molecules) did [51, 77, 197].

Similarly, the lithopanspermia experiments were flown at the same time as STONE 5 and STONE 6 [222]. In these experiments, microorganisms living naturally on rock surfaces, such as lichens and mosses, were exposed to space conditions. Although the BIOPAN experiments showed that they could survive in space, the heat of atmospheric entry was too high for their survival, even when they were protected by an extra layer of rock.

3.2.1.3 Future Perspectives for Astrobiology Research in LEO

Several new astrobiology experiments are in preparation for EXPOSE-R-2 (launch 2012) as well as innovative concepts to study radiation biology and prebiotic reactions inside and outside of the ISS. BIOMEX (Biology and Mars Experiment) is planned to obtain knowledge on stability and degradation levels of space-exposed pigments, secondary metabolites and cell surfaces in contact with a terrestrial and Martian analogue mineral environment.

BOSS (Biofilm Organisms Surfing Space) will test the hypothesis that the biofilm form of life with microorganisms embedded and aggregated in their extracellular polymeric substance (EPS) matrix is suited to supporting the long-term survival of microorganisms under the harsh environmental conditions in space and on Mars, and is superior to the same bacteria in the form of planktonic cultures.

The Miller-Urey Experiment in space (MUE) is investigating the formation of potential prebiotic organic compounds in the early solar system environment. PSS (Photochemistry on the Space Station) is proposing to use the EXPOSE-R facility for projects related to organic chemistry, and is accommodating a new experiment, to broaden the range of chemical compounds already exposed, in relation to astrochemistry and astrobiology.

VITRINE will be designed for long duration exposure in space (>1 year) with active temperature control, and the capacity to deal with samples at very low temperatures. A Sun-pointing device or equivalent mechanism (heliostat) will optimize the ratio of the exposure time to UV/duration in space environment. Instruments will be installed that monitor *in-situ* the evolution of the samples during the experiment's operation phase. Non-destructive analyses will be performed by infrared and ultraviolet spectroscopy, and destructive analyses will be performed by gas chromatography and/or mass spectroscopy.

OREOCUBE proposes the development of an integrated "single-cube" UV/visible/near-IR spectroscopy system for the *in-situ* study of the effects of the space environment on astrobiologically relevant materials as an outside exposure facility on the ISS.

A new series of STONE experiments are planned in which the emphasis will be on the fate of organic molecules associated with extraterrestrial-analogue materials (asteroid/Martian rocks) during atmospheric entry. These molecules will include the prebiotic components of the first cellular life, as well as certain molecular biomarkers.

In summary, future experiments are needed to investigate in more detail possible synergistic effects of orbital stresses on astrobiological samples including vacuum, UV radiation, ionizing radiation, thermal extremes and microgravity, to get more insights into space chemistry and to prepare effectively for human exploration.

3.2.2 Gravitational Cell Biology

Johannes Boonstra[1], Silvia Bradamante[2], Alain C. Colige[6], Ruth Hemmersbach[3], Charles A. Lambert[6], Natalie Leys[4], Jack J.W.A. van Loon[5], Betty V. Nusgens[6]

[1]Cell Biology, Department Biology, University Utrecht, Padualaan 8, 3584, CH Utrecht, The Netherlands
[2]CNR-ISTM, Institute of Molecular Science and Technology, Via Golgi 19, 20133 Milan, Italy
[3]Biomedical Science Support Center, Institute of Aerospace Medicine, German Aerospace Center (DLR), 51147 Cologne, Germany
[4]Unit for Microbiology, Institute for Environment, Health and Safety, Belgian Nuclear Research Center SCK/CEN, Boeretang 200, 2400 Mol, Belgium
[5]DESC @ OCB-ACTA, University of Amsterdam and VU-University, Gustav Mahlerlaan 3004, 1081 LA Amsterdam, The Netherlands
[6]Laboratory of Connective Tissues Biology, University of Liège, Liège, Belgium

3.2.2.1 Introduction

Organisms exposed to microgravity conditions experience various physiological changes. For example humans suffer from immunodeficiency, loss of bone mass and muscle deterioration that might be due to perturbations of mechano-signalling affecting cell homeostasis. The underlying mechanism for these phenomena might be within cells at the molecular level. The aim of our space cell biology programmes is, besides gaining basic knowledge, to understand the cellular and molecular mechanisms operating in the perception and transduction of mechanical signals, including gravity. An important goal is to disclose key pathways that might represent potential targets for pharmacological intervention to prevent microgravity-related health alterations as well as mechano-dependent diseases on Earth such as age-related osteoporosis.

Conceptually the effect of gravity on single cells is expected to be limited. Based upon a theoretical comparison of various forces within the cell it has been predicted very unlikely that non-specialized cells would display a response to changes in gravity [202]. However, at the same time numerous studies in space and on Earth have shown that *in vitro* cells do behave differently under conditions of modified gravity as compared to 1xg controls. Gravity may act on single cells directly via mass displacements of specific intracellular components or of the whole cell volume [158]. In addition, gravity may also act on cells indirectly by changed fluid dynamics via Rayleigh or density-driven convection of the surrounding media [158]. In this chapter we describe briefly the current status of the research performed, in a European context, on the effects of gravity on eukaryotic and prokaryotic cells and the underlying molecular mechanisms.

3.2.2.2 Effect of Microgravity on Human Epithelial Cells

Epidermal growth factor (EGF) plays an essential role in the proliferation of a wide variety of mammalian cells. Binding of EGF to its receptor activates signal transduction cascades leading to expression of specific genes including *c-fos* and *c-jun*. Studies during sounding rocket missions and in clinostats demonstrated that EGF-induced expression of *c-fos* and *c-jun* was decreased under microgravity conditions [for review see 25]. Furthermore it was demonstrated that protein kinase C (PKC)-mediated *c-fos* and *c-jun* expression was sensitive to microgravity conditions, in contrast to Ca^{2+} and protein kinase A (PKA)-mediated *c-fos* and *c-jun* expression (for review see [25]).

In addition cells showed increased cell rounding under microgravity conditions (for review see [25, 121]). Cell rounding is largely determined by the actin microfilament system. Interestingly, the relative F-actin content was shown to increase during microgravity conditions (for review see: [25, 121]). Microgravity therefore seems to induce polymerization of actin or reduce depolymerization of actin. This cell rounding was further studied in simulated microgravity using a random positioning machine (RPM) and by magnetic levitation. The RPM rotates randomly in all directions resulting in a net force of zero over a given period of time. In magnetic levitation a magnetic gradient is used to counterbalance the gravitational force, resulting in a stable levitation and the simulation of microgravity [16]. Both random rotation and magnetic levitation induced similar changes in the actin microfilament system that were also found under real microgravity conditions [179].

3.2.2.3 Involvement of the Rho Family GTPases in Mechano-Perception and Reaction, Including Gravity

The modifications of gene expression (increased expression of MMP-1 and IL-6) that were observed in human fibroblasts embedded in 3D-collagen gels, suggested that cells interpret microgravity as a mechanical relaxation. The current working hypothesis is that the reduced gravitational force experienced during space flight affects cell architecture, the Rho GTPases function and their cellular signalling. The Rho GTPases are key regulators in the organization and turnover of the cytoskeleton (CSK), the formation and signalling function of cell-matrix contacts (focal adhesions, FA) and the control of cellular pathways regulating cell survival, proliferation, gene expression and mechanical functions such as adhesion, migration and contraction [190].

To evaluate the involvement of the Rho GTPases in the mechano-perception and –reaction to microgravity, a double strategy was used: (i) inactivating Rho GTPases by small interfering (si)RNA (siRhoA, siRac1, siCdc42) or (ii) forcing the expression of a constitutively active mutant of each of these Rho GTPases (RhoAQL, Rac1QL, Cdc42QL) in fibroblasts (WI26) and osteoblasts (MG63) [230, 275]. The end-point analysis evaluates the architectural organization of the CSK and FA by computerized image analysis, the migratory capacities and the transcriptomic repertoire of the engineered cells under microgravity and 1 g conditions by comparison with the wild type cells. An attenuation or exacerbation of the microgravity effects in these genetically engineered cells will make it possible to identify the upstream and downstream Rho GTPase(s)-dependent signalling pathways disturbed in microgravity. The complete series of wild type and mutant cells was used in several missions generating a considerable amount of data that are still under analysis.

3.2.2.4 Effect of Microgravity on Growth and Morphology of Endothelial Cells

Vascular endothelial cells (ECs) play an important role in tissue homeostasis, blood-tissue exchange and vasotonic regulation and are very sensitive to the physical environment. Dysfunctions in macro/microvascular ECs cultured in simulated microgravity have been described [169, 256]. In particular, simulated microgravity stimulates Human Umbilical Vein EC (HUVEC) growth and this effect correlates with an over-expression of hsp70 and a down-regulation of IL-1α, a potent inhibitor of EC growth, also implicated in promoting senescence. In addition, HUVECs rapidly remodel their cytoskeleton, markedly down-regulate actin, and increase the synthesis of vasodilators with significant modulation of gene expression. Simulated microgravity influences HUVEC proliferation and differentiation with the involvement of anti-angiogenic factors that may be responsible for the non-spontaneous formation of blood vessels. The results of the simulations were validated with an experiment in space, specifically aiming at the evaluation of EC gene expression, protein profile, and synthesis of nitric oxide. The analyses of the space-flown cells are currently ongoing.

3.2.2.5 Effect of Microgravity on Proliferation, Differentiation and Morphology of Bone Cells

One of the main physiological impairments observed during long-duration space missions is a progressive loss of bone density strongly resembling osteoporosis. Considering that Mesenchymal Stem Cells (MSCs) contribute to healthy bone, understanding their biological response to microgravity will help to identify the mechanisms of some of the physiological changes occurring during spaceflight. It was demonstrated that MSCs from different origins (human and goat) cultured under simulated microgravity conditions modulate their proliferation and differentiation depending on the culture medium [257]. In detail, a decrease was observed in human MSC proliferation with low osteoblast differentiation in the RPM under osteogenic stimulation. This is possibly due to either (a) inhibition of cell growth, (b) cell death, (c) differentiation in adipocytes, or (d) a combination of effects.

It is known that the mechano-adaptation of bone involves osteoblasts and osteoclasts as well as bone matrix cells, osteocytes. Signal transduction molecules such as prostaglandins, as well as mechano-sensitive cation channels, G-protein-dependant pathways, protein kinase C and nitric oxide are involved in mechanosensing (for review see [255]). It has been hypothesized that the impact of gravity on bone cells may result in small changes in cell shape. To explore such small changes, cell height was determined at various gravitational loads using an Atomic Force Microscope (AFM) operated in liquid [156, 159]. A monolayer of an osteoblastic cell line, MC-3 T3-E1, was cultured to subconfluency and placed in the centrifuge-AFM system exposing it to 1 g (non-rotating), 2 g and 3 g. The pilot experiment revealed the possible operation of an AFM under hypergravity conditions and that, in real time, bone cells do indeed adapt their shape in response to an increased gravitational load from only 1 or 2 g.

3.2.2.6 Graviperception in Unicellular Protists

Protists are free-living unicellular organisms which need sensors for environmental stimuli in order to survive and to actively reach optimal living zones. As gravity is the most reliable cue, being the only natural constant, a sensor for using gravity with respect to spatial orientation (gravitaxis) and movement speed (gravikinesis) will contribute to survival even under unfavourable living conditions. Studies on gravisensitive protists help to identify gravity-related signal transduction pathways. Already at the beginning of the nineteenth century there were enthusiastic discussions of the observed negative gravitaxis (upward swimming) of some protozoan species, considering it either as a purely physical (buoy effect) or physiologically guided mechanism (for review see [99, 103]). Various experiments in microgravity on sounding rockets and parabolic flights, as well as extensive ground-based studies, confirmed that gravity perception in protists is a physiologically guided mechanism supported by physical and geometric properties of the cell. While in the ciliate *Paramecium* and the flagellate *Euglena* the "heavy" mass, whose displacement is the prerequisite for initiating a sensory process, is the mass of the cell itself (density 1.02–1.04 g/cm^2), representatives of the family Loxodidae contain distinct cell organelles for graviperception (Müller organelles). As a conclusion, graviperception processes can be reduced to two principles: perception via the whole protoplast or via intracellular statoliths [99, 103]. Behavioural studies in microgravity – using computer-based image analysis – revealed the role of gravity in regulation of orientation and speed, demonstrated the fast responsiveness of gravisensitive protists, their lack of behavioural adaptation (at least in the 14-day time frame) and the existence of thresholds in the range of 0.12–0.32 g. Electrophysiological studies demonstrated the existence of gravireceptor potentials based on the activation of mechano-sensitive ion channels in the corresponding lower cell membrane (depending on the orientation of the cell) [135]. Loss of gravitaxis after lesion of the Müller organelles of *Loxodes* proved their gravisensory function. Recent studies have concentrated on the gravity signalling pathway in *Euglena* using online monitoring in microgravity as well as molecular biological approaches. Activation of mechano-sensitive ion channels due to an altered load on the lower membrane increases Ca^{2+} inward gating, an increase in the intracellular Ca^{2+} concentration, and in turn activation of one of the several calmodulins, followed by activation of an adenylyl cyclase, finally resulting in cAMP production [58, 244]. It is assumed that cAMP activates a specific protein kinase A which in turn regulates the flagellar beating mechanisms by protein phosphorylation and, finally reorientation of the cell.

3.2.2.7 Effect of Microgravity on Yeast

Exposure to the space environment has been associated with an increase in oxidative damage in yeast cells which is possibly due to the generation of high-energy free radicals. The oxidative stress response, *i.e.* the alterations in glutathione (GSH) homeostasis in particular, of the yeast *Saccharomyces cerevisiae* was studied under real microgravity conditions during space flight. Analysis of the space-flown samples showed that the yeast responded to the imposed stress (microgravity and oxygen) by activating signal transduction cascades resulting in the extracellular release (ER) of GSH. Also under simulated microgravity conditions *S. cerevisiae* exposed to oxidative stress released significant amounts of endogenous GSH in the culture medium. The results, validated on the ground, indicate the role of actin in GSH ER. It was found that the GSH ER can be inhibited using a potent chloride channel blocker, demonstrating that the actin-related activation of chloride channels is the main process responsible for GSH release [33]. These results significantly contribute to a better understanding of the mechanisms affecting GSH homeostasis in organisms, and may provide insights into the underlying cause of GSH depletion in many diseases.

3.2.2.8 The Impact of Gravity on Bacteria

The way bacteria respond to gravity is still unclear. Early studies suggested that their uniform density and small size (<10 μm) are theoretically unlikely to enable direct sensing of gravity [202]. Also more recent studies have suggested that bacterial cells may rather indirectly respond to reduced gravity conditions, because of changes in their immediate environment. Below 1 g, one can observe buoyant flows in cultures of non-motile bacteria, originating from solute gradients created by cell metabolism; very similar to solute-induced buoyant flows around growing protein crystals [15]. Reduced gravity suppresses buoyancy-driven convection and thus limits the mechanism of mixing of fluid in bacterial cultures to diffusion [2]. How exactly a given microbe in a microbial community will adapt to and behave in space, however, depends on many factors and is still difficult to predict. Several space flight experiments and simulation experiments on Earth have shown that different bacteria can differ significantly in their responses to microgravity. Moreover the response of a bacterial cell to changes in gravity is also dependent on its status and matrix (*e.g.* dormant/actively growing, suspended or on the surface, in rich or minimal medium, in light or dark, *etc.*). Nevertheless, in general one could conclude that real microgravity has been shown not to hinder bacterial proliferation, on the contrary, it may enhance the growth rate of some bacteria [133]. Gene and protein expression studies performed on a variety of bacteria (*e.g. Cupriavidus, Pseudomonas, Rhodospirillum, Escherichia, Salmonella, etc.)* under both space and simulated microgravity conditions support the hypothesis that bacteria actively respond to the fluid quiescence (low shear forces) and/or the changes in nutrient availability and metabolite removal imposed by the altered mass transport under these conditions [54, 153, 154, 171, 261]. One working hypothesis is that the high growth rate and the creation of zones of nutrient depletion over time in their immediate surroundings make these bacteria respond in a way that is similar to starvation. Moreover, microgravity was shown to induce (potentially through this 'starvation-like' signal) several metabolic and physiological changes, including increased synthesis and secretion of certain metabolites, higher resistance to stresses such as heat, acids, salt, oxidants and antibiotics, and even an increased virulence in some pathogenic bacteria.

Acknowledgments The research described in this chapter has been supported by: the Dutch Space Organization (NWO-MG-057, MG-059); EuroMagNET; European Space Agency; Italian Space Agency (ASI); German Aerospace Center; the Belgian Science Policy (BELSPO); ESA/PRODEX Office; the Belgian Fonds National de la Recherche Scientifique (FNRS); ESA-GSTP EC-FP7 program

3.2.3 Radiation Biology

Günther Reitz[1], Marco Durante[2]

[1]DLR, Radiation Biology, Cologne, Germany
[2]GSI, Biophysics, Darmstadt, Germany

3.2.3.1 Radiation Field in Deep Space

Current space programs are shifting toward planetary exploration and in particular towards human missions to the Moon and Mars. However, space radiation is a major barrier to human exploration of the solar system because the biological effects of high-energy and charge (HZE) ions, which are the main contributors to radiation risks in deep space, are poorly understood. Predictions of the nature and magnitude of the risks posed by space radiation are subject to large uncertainties. Great efforts have been dedicated worldwide in recent years toward a better understanding of the biological effects (including both cancer and noncancer effects) of galactic cosmic rays. Most of the annual effective dose for terrestrial populations comes from α-particles (from radon in air), γ- and β-rays (from natural internal and terrestrial sources), and X-rays (from diagnostic medicine). In contrast, in space most exposure of astronauts inside spacecrafts in Low Earth Orbit (LEO) is incurred by trapped protons in the Earth's radiation belts with energies below 300 MeV and by particles from galactic cosmic radiation (GCR) with energies up to several GeV per nucleon (Fig. 3.27). Whilst in LEO an additional shielding effect

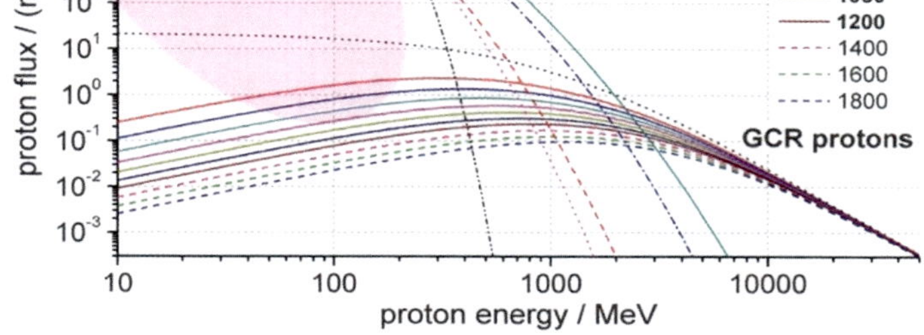

Fig. 3.27 Particle spectra for galactic cosmic radiation protons for different times in the solar cycle (typical solar minimum occurs at values starting with 300 MV of the heliocentric potential, solar maximum up to 1,200 MV) together with particle spectra from some major solar particle events presented as Weibull functions. The purple range represents the protons in the radiation belts

arises from the deflection of charged particles by the Lorentz force, in deep space GCR of all energies is present.

GCR is present isotropically in space, and its intensity is modulated by solar activity: the GCR flux is indeed approximately 2.5 times higher at solar minimum than at solar maximum. The energy spectrum of all GCR particles is very wide, and peaks between 0.3 and 2 GeV/n. Figure 3.27 shows the solar modulation for the proton component. Although elements from $Z = 1$ (H) to $Z = 92$ (U) are present in the GCR, ions heavier than iron are very rare, and heavier particles only represent 1% of the total GCR flux. Nevertheless, because the absorbed dose is proportional to Z^2 and the radiobiological efficiency is very high for heavy ions, protons contribute less than 20% to the radiation exposure, and iron alone gives approximately the same contribution (Fig. 3.28).

In addition to chronic GCR exposure, acute doses may be caused by large solar particle events (SPEs), which consist of very large fluxes of protons at energies that are usually below 200 MeV, but that occasionally may reach the GeV region (see spectra given in Fig. 3.27). For astronauts in LEO, the geomagnetic field provides efficient shielding against these particles, but during an interplanetary mission or on the planet's surface, crewmembers may be exposed to a lethal proton dose if not adequately protected. SPEs are unpredictable, but are known to be more frequent at solar maximum when the GCR flux is reduced by the Solar magnetic field. Hence, interplanetary missions that occur during solar maximum will result in reduced exposure to GCR, although an increase in risk could arise from exposure to a large SPE [208].

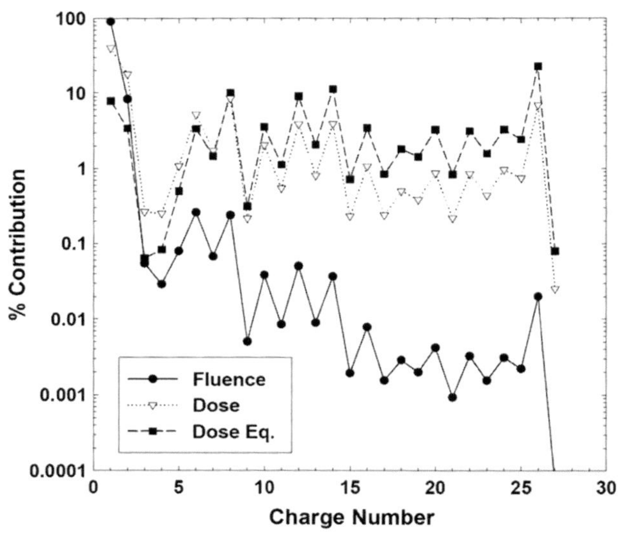

Fig. 3.28 Relative contribution of different ions to flux (fluence), dose, and dose equivalent from galactic cosmic radiation. Nuclei heavier than nickel are very rare. Calculation by HZETRN kindly provided by Dr. Frank Cucinotta and Dr. Cary Zeitlin

3.2.3.2 Risk Uncertainty

No effective countermeasures against space radiation exposure are currently available. Shielding is effective against SPE particles, especially for low-energy events, but it is unsuitable for the very energetic GCR that produces secondary particle showers in target materials by nuclear fragmentation. Even bulky and heavy aluminium shielding can reduce the GCR effective dose by no more than 30%. Light, hydrogen-rich materials, such as polyethylene, are more effective shielding materials, but the reduction in GCR dose will be unlikely to exceed 50%. Medical countermeasures, such as radioprotective drugs or dietary antioxidants, are currently still in development.

Clearly, it will be impossible to reduce exposure by shielding to the levels found on Earth. This raises the question of what dose is acceptable to maintain safety. Doses in the Apollo lunar exploration program ranged from 10 to 30 mSv, and the Apollo 16 crew luckily missed being exposed to the large SPE in August 1972 by only four months. Projected exposures in different mission scenarios have been evaluated using radiation transport codes. Exposures in a Martian mission of 1,000 days may result in a dose equivalent of up to 1,000 mSv for GCR only, whereas a worst case SPE like the Carrington flare can deliver this dose in one single exposure. Using a procedure similar to the standards adopted for workers occupationally exposed to radiation on Earth, the US National Council for Radiological Protection (NCRP) has recommended career dose limits for astronauts involved in LEO missions [187]. There have been until now no dose limits for interplanetary missions, simply because the uncertainty in risk estimates for the exposure to GCR is far too high [188]. Most of the uncertainty in risk is related to the radiation quality of heavy ions, to the dose and dose-rate effectiveness factor (DDREF) to be used for cancer risk estimates, and to tissue degenerative effects, such as central nervous system (CNS) damage, which appear to be unique to heavy ion exposure [62]. Uncertainty in estimates for increased cancer risk from a Mars mission is currently between 400% and 600% (Fig. 3.29). Uncertainty for non-cancer risks is presumably even higher.

3.2.3.3 Heavy Ion Radiobiology

The main biological effects associated with exposure to cosmic radiation are [188]:

- Carcinogenesis
- Late degenerative tissue effects
- Acute effects
- Hereditary effects

Cancer is currently estimated to dominate risk estimates, and the LEO dose limits are dominated by cancer mortality risk. However, non-cancer effects are becoming an increasing source of concern. They can further be divided into:

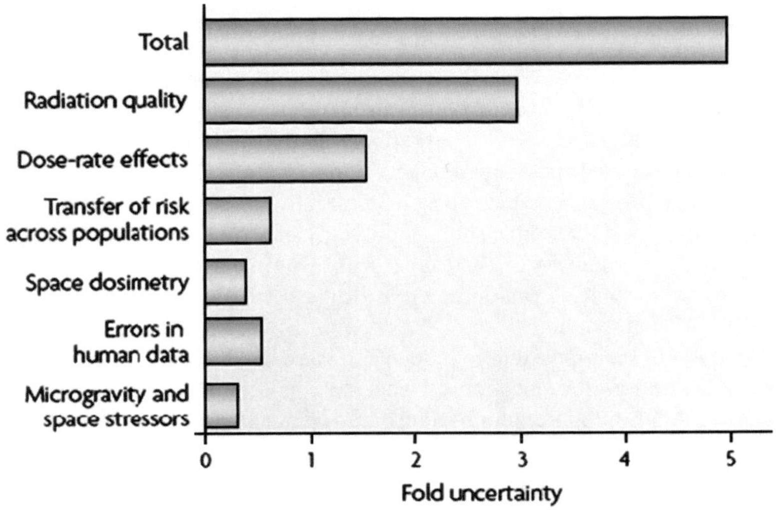

Fig. 3.29 Estimates of uncertainties in projecting cancer risks for space and terrestrial exposures. Several factors, such as radiation quality effects, space physics, and microgravity do not apply on Earth and lead to large increases in risk projections. Predicting risks to individuals is difficult as there are very few quantitative measures of individual sensitivity. Only a select few individuals enjoy space travel and projecting risks for a few selected individuals rather than populations will be of utmost importance for space missions to Mars. The extrapolation from experimental models to humans is perhaps the greatest challenge to cancer risk assessments. Plot form ref. [62]

- acute and late damage to the central nervous system (CNS);
- cataract formation;
- cardiovascular diseases including coronary heart disease and stroke;
- digestive and respiratory diseases;
- accelerated senescence leading to endocrine and immune system dysfunction.

Acute effects are expected only for very intense SPE exposure. The prodromal signs such as nausea, vomiting, and fatigue are the most likely, and acute mortality is very improbable. Hereditary effects are generally assumed to be a factor of 10 or lower than somatic effects in radiation protection, although there is a lack of evidence for heavy ions in this field. The age at flight and the small number of astronauts tends to reduce the importance of hereditary and reproductive effects.

Interestingly, so far cataract is the only cosmic radiation-induced effect actually observed in astronauts [44]. Recent epidemiological evidence does not support the existence of a dose threshold for radiation-induced lens opacification, although the shape of the dose-response curve is still unclear. However, animal studies show that the relative biological effectiveness (RBE) of heavy ions for cataract induction is as high as 50 at doses below 100 mGy, thus explaining the high effectiveness for accelerated cataractogenesis in space [20].

The RBE of HZE particles compared to X-rays represents a major source of uncertainty (Fig. 3.29), and indeed the problem of radiation quality is central to radiation protection on Earth as well. The shape of the dose-response curve at low doses is a key problem, and the importance of non-targeted effects or adaptive response mechanisms is still unclear both for radiation protection on Earth and in space. Conventional risk assessment approaches have been limited to using averaged quantities such as dose or dose equivalent.

The large number of health risks of concern, entailing many tissues and diseases, complicates pursuing approaches such as genetic selection, because it is highly unlikely that one person could be resistant to so many diseases. Approaches to biological countermeasures are also challenged by the multitude of health risks as well as the types of radiation in space and the chronic exposure of up to 3 years for a Mars mission. These factors limit the applicability of radiation protection measures developed for terrestrial exposures, which are usually for acute exposures to low LET radiation.

Large ground-based radiobiology research experimental programs are currently ongoing to reduce uncertainty regarding the biological effects of heavy ions, both in the USA (at the NASA Space Radiation Laboratory in the Brookhaven National Laboratory, NY), Europe (at GSI in Darmstadt, Germany), and Japan (at HIMAC in Chiba). Results from these experiments have been reviewed in several recent publications [56, 62, 63, 188], and the reader will find all the most up-to-date biological information in those papers.

Acknowledgments We thank DLR and ESA (IBER-contract) for their support to space radiation research in our Institutes

3.2.4 Plant Sciences

F. Javier Medina[1], Stefano Mancuso[2], and Rüdiger Hampp[3]

[1]Centro de Investigaciones Biológicas (CSIC), Ramiro de Maeztu 9, 28040 Madrid, Spain
[2]LINV – Department of Plant, Soil & Environmental Science, University of Florence, Viale delle idee 30, 50019 Sesto Fiorentino, FI, Italy
[3]University of Tübingen, Microbiology Institute, Auf der Morgenstelle 1, D-72076 Tübingen, Germany

3.2.4.1 Plant Gravity Sensing and Gravitropism

Sensors of Gravity

Weightlessness is a key difference between the space and terrestrial environments. Plant growth is conspicuously dependent on the direction of the gravity vector, since plants continuously check their growth direction according to this vector in a process called gravitropism. This results in roots growing downwards and shoots growing upwards. Plants have implemented a complex mechanism of gravity sensing, whose purpose is to convert physical information into physiological (chemical) information, capable of being transduced and integrated in the general plant growth machinery. The roots and shoots of higher plants have cells specialized in gravity sensing called statocytes; these cells contain specialized plastids, containing large starch grains (amyloplasts) called statoliths. According to the classical "starch-statolith theory", gravity susception is based on the movement of statoliths, sedimenting in the direction of gravity [199]. This theory has been reinforced in different plant species as representing the major mechanism of gravity susception, but alternative mechanisms may coexist, such as the hydrostatic pressure model [240].

In roots, the sensor organ is the columella, located in the root cap. Indeed, physical or genetic ablation of the root cap results in loss of gravitropism. There is a secondary sensor of gravity in the transition zone of the root, where gravity is sensed by the hydrostatic pressure mechanism. In shoots of dicotyledonous plants, gravity is sensed in hypocotyl endodermal cells [240].

Statolith movement triggers the activation of mechano-sensitive ion (Ca^{2+}) channels [70]. The statolith's outer membrane makes physical contact with membranes of the endoplasmic reticulum (ER) arranged at the periphery of the statocytes. Since ER is an intracellular Ca^{2+} reservoir, the contact of the two membranes produces the release of the stored Ca^{2+} [151]. Activation of ion channels in membranes has been also proposed to be a consequence of the distortion of the actin microfilament network that surrounds amyloplasts and connects them with the plasmalemma, owing to the movement of these organelles [235]. Repeated short-term gravistimulation, for periods not sufficient for sedimenting amyloplasts to reach the bottom of the cell, but sufficient to alter the surrounding actin cytoskeleton, was enough for inducing a gravitropic response [199]. However, depolymerization of actin filaments increases the response to gravity, thus supporting an opposite view, namely that statocytes are sensitive to gravity because their cytoskeleton is less robust and more elastic [7]. Interestingly, decapped roots regain their ability to perform root gravitropism if they are devoid of actin filaments, confirming that there are graviresponsive tissues in the root other than the root cap [163].

As an alternative or simultaneous mechanism, statolith movement may cause contact between the outer amyloplast envelope and the plasma membrane (or any other endomembrane), triggering a ligand-receptor interaction in which the receptor would act as a signal transducer capable of propagating the mechano-signal [240]. Interestingly, the discovery of proteins playing a role in ligand-receptor interactions in the vesicle trafficking apparatus may serve to connect the two models proposed [240]. In fact, this apparatus is associated with the cytoskeleton, since endo-, exocytosis and vesicle movements are all dependent on actin filaments and myosin.

Transduction of Gravistimulus

Once the information coming from the gravistimulus is converted from physical to physiological in the statocytes, it is transmitted to the neighboring cells, leading, eventually, to a relocation of the auxin efflux carriers responsible for the lateral transport of auxin. Asymmetry of the auxin concentration causes a differential growth rate that, in the case of the roots, is initiated in the transition zone [8]. The subsequent growth increase in the upper half and growth inhibition in the lower half results in root curvature. How this process of conversion and transmission occurs is a question that remains to be fully clarified.

Many signalling molecules, such as Ca^{2+}, have been detected experimentally in response to gravistimulation. Recently, auxin-induced NO and cGMP have been added to the list of possible mediators in the primary root of soybean [120]. Reactive oxygen species (ROS) have also been shown as downstream constituents in the auxin-mediated signal transduction in root gravitropism. H_2O_2 is asymmetrically generated in roots after gravistimulation, as it is with asymmetrically applied auxin. Moreover, ROS scavengers inhibit the gravitropic response [125].

3.2.4.2 Gravity Response in Higher Plants

Molecular Level: Alterations in Gene and Protein Expression

Gene Expression

The availability of gene arrays has stimulated the investigation of gravitational effects on gene expression. Early gravity-related studies using Arabidopsis as a model plant were published (see for example [4, 167, 184]). A wide range of gravity-regulated genes were identified. They belong to functional groups involved in cellular organization and cell wall formation / rearrangement, signalling, phosphorylation/ dephosphorylation, proteolysis and transport, hormone synthesis plus related events, defence, stress-response and gravisensing. Many of the alterations are part of a general stress response, but changes related to the synthesis / rearrangement of cell wall components are probably hyper-g specific. They also indicate a possible interplay between gravitropic and other mechanical responses. With regard to metabolism, hypergravity effects indicate a possible shift from starch synthesis to degradation, and increased anaplerotic activities (PEP carboxylase: ketoacid supply for amino acid synthesis; stress response) and secondary metabolism [167].

Protein Expression and Modulation

Changes in protein expression were studied by transferring seedlings from the vertical to the horizontal position and watching root curvature for up to three hours. A proteomic analysis identified ten proteins whose amounts decreased, while six were increased by gravitational stimulation [127]. In another proteomic study, Arabidopsis callus cultures were subjected to horizontal and vertical clinorotation for 8 h. Significantly altered expression was found in 18 proteins. Of these, seven were involved in stress responses, and four were identified as key enzymes of carbohydrate and lipid metabolism, or could be related to cell wall modifications [265].

By exposing *Arabidopsis* cell cultures to 8 *g* (centrifugation) and simulated microgravity for up to 16 h, 28 spots were found to change in amount after 2 h of hypergravity (18 up-, 10 down-regulated). The identified proteins were stress-related, and are involved in detoxification of ROS, signalling, and calcium binding [11].

Furthermore, from sounding rocket experiments, there is evidence for metabolic alterations during short-term exposure to μ*g* (up to 6 min) [101]. Cellular responses mainly dwell on protein modulation (phosphorylation / dephosphorylation) and on the initiation of signalling cascades. Exposure of *Arabidopsis* cell cultures to hypergravity (8 *g*) and simulated near-weightlessness for up to 30 min resulted in 18 (hypergravity) and 25 protein spots (sim. μ*g*) with increased/decreased signal intensity of at least twofold [12]. Out of these, 36 could be identified. The respective proteins are involved in ROS scavenging and detoxification, primary metabolism, general signalling, protein translation and proteolysis, and ion homeostasis. These findings are further evidence that gravity changes induce ROS production (see also Fig. 3.30).

With regard to microgravity, considerable changes in protein phosphorylation occur after only 20 s (parabolic flight data, Hampp et al. unpublished).

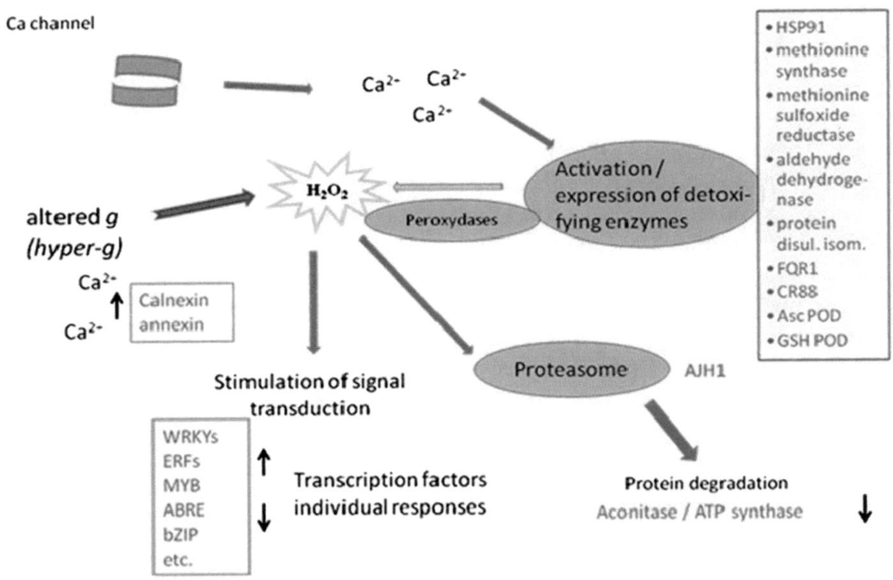

Fig. 3.30 Summary of potential hypergravity-gravity dependent signal transduction chains in a plant cell (*Arabidopsis thaliana*). Alterations in the gravitational field increase intracellular Ca^{2+}-concentration. In parallel, reactive oxygen species (ROS) are released, which cause different reactions in the plant cell. (Formation H_2O_2 was observed after exposure of A.t. cell cultures to 7 g for 30 min; Hampp, unpublished). Due to ROS production, antioxidant enzymes are expressed or activated (see box). Furthermore ROS lead to the enzymatic activation of proteasomes that foster protein degradation. Additionally, ROS can mobilize proteins called annexins, which are able to interact with the second messenger, Ca^{2+}. ROS can also interfere with gene expression

Cellular Level: Alterations in Cell Structure and Functions

Altered gravity not only affects sensor cells, but also features of cells not specialized in gravity perception. These include stem cells in the root meristem that play a key role in developmental events. The function of meristematic cells consists essentially of the consecutive activities of cell division and cell growth, which are organized in the highly regulated mechanisms of the cell cycle. In this context, meristematic cell growth means exclusively the increase in cell size which allows cell division mostly consisting of protein synthesis [178].

In order to determine the effects of weightlessness on highly proliferating root meristematic cells, young seedlings were analyzed for parameters of cell proliferation and cell growth (nucleolar parameters). These seedlings originated from seeds germinated in space or in ground facilities providing a functional simulation of microgravity, such as the Random Positioning Machine, RPM [157]. Seedling culture in microgravity resulted in an enhancement of cell proliferation and a depletion of cell growth. Therefore, these two essential cellular functions, which are closely coordinated under normal ground gravity conditions, lost their coordination and appeared uncoupled [168] (see also Fig. 3.31).

It is known that auxin coordinates cell growth and proliferation. Auxin stimulates cell division and cell cycle progression and low auxin levels induce cell expansion and differentiation [200]. Furthermore, auxin is a mediator of the

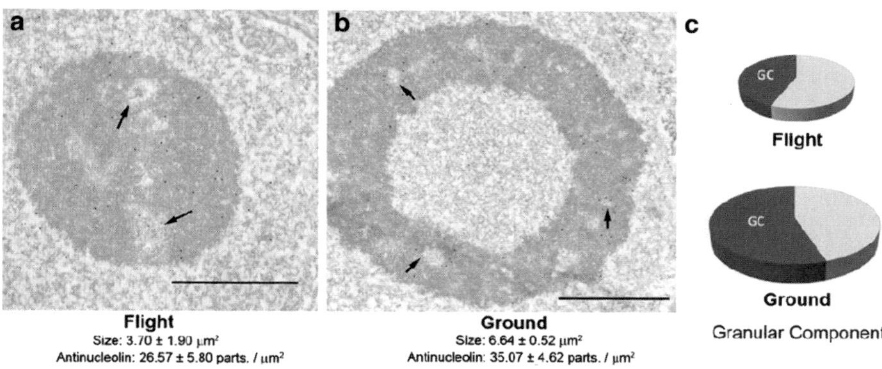

Fig. 3.31 Estimation of the alterations in root meristematic cell growth after spaceflight obtained from structural, morphometric and immunocytochemical data of the nucleolus. Ultrastructural images of nucleoli from 4 day-old seedlings grown in the International Space Station (**a**) and, in parallel, on Earth (**b**) showing the immunogold localization of the nucleolar protein nucleolin. Nucleoli from the space-grown sample are smaller and have lower levels of nucleolin than the ground control. Note that the distribution of nucleolin is concentrated around fibrillar centers (*arrows*) in the ground control sample (**b**), but gold particles appear more dispersed throughout the nucleolus in the flight sample (**a**). The proportion of granular component (*GC*) is lower in the flight sample than in the ground control (**c**). All these parameters indicate a reduction in the rate of ribosome biogenesis produced in weightlessness and, consequently, a lower level of protein synthesis, which, in these cells, means a reduction of cell growth

gravitropic signal transduction; thus, changes in the distribution of auxin in roots, caused by the transduction of an altered gravistimulus, not only result in the change of direction of root growth but may also produce changes in the expression of specific growth coordinators, leading to an alteration of the cell cycle and protein synthesis [172].

Microgravity as a Stress Factor for Plants: Alterations in Development and Physiology

Alterations in plant growth and physiology under microgravity occur at any level from subcellular to developmental, producing classical symptoms of plant stress. Investigations of plant growth and development in microgravity have primarily focused on gravity-dependent plant processes such as gravitropism, but attention must also be paid to distinguishing alterations due to microgravity from secondary multifactorial stress effects (gas-exchange, light, radiation) caused by the perturbation of gravity-mediated physical processes. Many spaceflight experiments have reported changes in plant structure and metabolism consistent with the onset of symptoms of oxygen deficiency [243]. Under these conditions, plant metabolism characteristics are changed due to low oxygen availability, including the accelerated and preferential synthesis of glycolytic and fermentative enzymes [162], the accumulation of carbohydrates, and stomatal closure in the leaves.

The developmental aspects related to microgravity hardly interfere with reproduction. In the last 30 years, a few species have been grown from seed to seed in microgravity, showing that the plant life cycle does not necessarily require normal gravity (1 *g*) for its completion. Despite this, the quality of the organs and of the whole plants developed in low gravity is somewhat different, affecting, for example, the seed reserve lipid utilization [35], or pollen development [136].

In general, regarding plant survival in microgravity, the results reported in the literature are often contradictory, depending on the plant material and on the experimental conditions. However, a common result emerging from many different experimental approaches is that plants are able to conclude satisfactory the reproductive cycle from seed to seed, exhibiting, once more, an extraordinary capacity to adapt their body and physiology even to stresses that have never occurred to plants during the process of evolution.

Acknowledgments This work was supported by grants from the Spanish Plan Nacional de I + D+i (AYA2009-07952) (FJM), from Ente Cassa di Risparmio di Firenze (SM), and from Deutsches Zentrum für Luft- und Raumfahrt (RH)

3.2.5 *Animal Physiology*

Eberhard Horn[1], Jack J.W.A. van Loon[2], Jessica Aceto[3], Marc Muller[3]

[1]Karlsruhe Institute of Technology (KIT), 76128 Karlsruhe, Germany
[2]DESC, ACTA-OCB, VU-University and University of Amsterdam, Amsterdam, The Netherlands
[3]LBMGG, University, 4000 Liège, Belgium

3.2.5.1 Introduction

The study of physiological processes in animals using space conditions has two aims: (1) to support human physiology in those questions that cannot be studied in humans for ethical reasons, using suitable animal models, in particular for developing countermeasures for spaceflight-induced disturbances; and (2) to analyse the basic mechanisms of adaptation to space conditions in adults and during development. The complex degree of interactions between the various organs for a stable working organism favours the use of only a few standard animal models such as rats or mice, amphibians (*Xenopus* or *Pleurodeles*) or fish (zebrafish or Medaka fish) that are closely related to men; this allows an insight to be gained into "vertebrate" adaptation mechanisms to the space environment. However, identifying general mechanisms of adaptation to space requires the use of a variety of animal species including invertebrates such as fruit flies, nematodes, crickets, or even scorpions, rotifers and sea urchins (comparative approach). A requirement for physiological studies with animals is a suitable survival system since each species prefers specific environmental conditions and nutrition. Although low-cost solutions are available [117, 218], there remains a necessity for species-specific hardware adaptations. This fact may favour the use of standard vertebrate models but would then entail the loss of significant information about *basic* mechanisms of adaptation.

3.2.5.2 The Actual Status of Animal Physiology in Space Life Sciences

During the past decade, studies about the impact of microgravity on animal physiology have experienced a significant drop, that has been accompanied by the increasing attractiveness of studies in cell biology, molecular biology and genetics. Consequently, the number of studies using classical techniques such as neurophysiology or biochemical approaches to the study of metabolism has decreased. For reports on space-flight animal studies such as analysis of the metabolism of motoneurons or the sensorimotor cortex of rats (cf. [89]) we still very much rely on the data from the Russian Cosmos flights from earlier times.

Lack of sufficient space-flight opportunities has led to an increase in the number of experiments using *simulated microgravity* by means of machines such as the clinostat, Rotating Wall Vessel (RWV; [134]), Random Positioning Machine

(RPM; [26]), or Magnetic Levitation [16], or by means of animal-linked approaches such as tail or hind limb suspension. Parabolic flights and Drop Towers support studies of short-duration physiological events but not of adaptation to long-term space exposure. The RWV and RPM techniques mainly considered cell cultures and tissue engineering. Successful applications of RWV and RPM to physiology are rare; studies have included functional vestibular development in zebrafish *(Danio rerio)* [181], gene expression in a variety of developing organ systems in live embryos of transgenic zebrafish [232, 233], and cultivation of complete mouse organs [100]. As a complement to microgravity experiments, various studies have also been performed in hypergravity on centrifuges on the ground, often resulting in physiological effects that are the opposite of those seen in microgravity [106, 263]. The predictive power of hypergravity experiments for responses to space flight was therefore proposed [263]. However, genes such as *hsp70* were shown to be related to stress by both abnormal gravity conditions [232] while other genes were influenced by only one condition [106]. The relation between these gene regulatory events and the physiological outcome is at present unclear.

Tail suspension is a simulation technique with a high potential to detect physiological changes induced by un-loading. The number of tail suspension studies about orthostatic intolerance, and cardiovascular, sensorimotor and neuronal or immunological dysfunctions has increased steadily since this technique was cited for the first time by PUBMED in 1979 (Fig. 3.32). However, the years since 2006 have been characterized by a dramatic drop in physiological but not molecular studies.

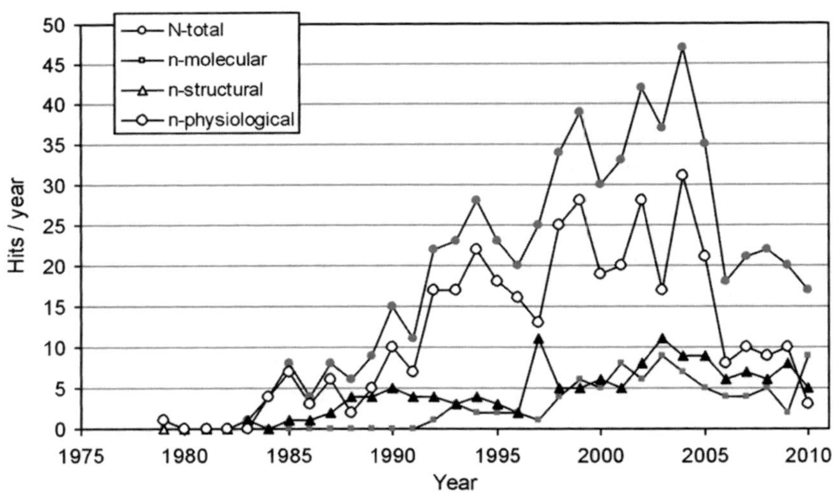

Fig. 3.32 Total number of articles per year using PUBMED and the keywords TAIL SUSPENSION and MICROGRAVITY (N-total). In addition, the total number has been subdivided into numbers of publications dealing mainly with physiological, structural or molecular questions

3.2.5.3 Physiological Adaptation to Microgravity

Adaptation to modified environmental conditions is a continuous process over time. Knowledge about patterns of time kinetics concerning adaptation to microgravity is low. Existing hypothetical descriptions have been derived from spaced but not from mandatory continuous physiological recordings, and mainly present saturation characteristic curves (cf. [3]). They overlook the possibility that adaptation might follow an "oscillatory pattern" in which ups and downs alternate in a damped manner to reach a stable physiological level. The oscillation type is not unlikely as shown by recordings from utricular afferents of toadfish (*Opsanus tau*) after exposure to hypergravity [29] or by in-microgravity recordings from the eighth nerve of bull frogs [31]. However, the studies so far reveal some well-founded basic aspects about mechanisms of adaptation to microgravity and readaptation to 1 g-Earth conditions; this information mainly results from comparative animal studies.

Vestibular sensitization. Anatomical, molecular, neurophysiological, behavioural and psychophysical studies in snails, fish, amphibians, rats and astronauts have revealed that the *sensitization* of gravity-sensing systems is a mechanism of adaption to long-duration space weightlessness. In rats, a 16-day space flight caused a tonic depression of the number of Fos- and FRA-labelled cells in efferent [5] but not in afferent vestibular nuclei [204]. The influence of the depressed efferent activity increases the activity flow from the peripheral vestibular structures indicating vestibular sensitization during exposure to microgravity. In fish (*Oreochromis mossambicus*) [227] and toads (*Xenopus laevis*) [114], the roll-induced vestibulo-ocular reflex was augmented for some developmental stages after termination of space weightlessness (Fig. 3.33), whereas it was decreased after exposure to 3 g [113].

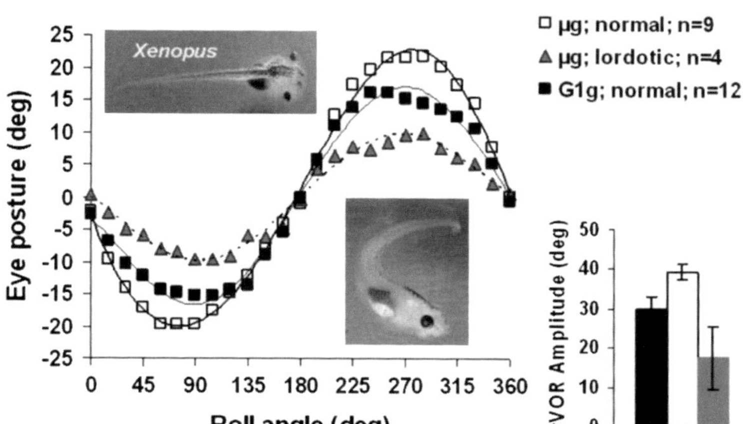

Fig. 3.33 Augmentation of the roll-induced vestibulo-ocular reflex (rVOR) after a 9.5-day space flight (Andromède 2001) in normally developed *Xenopus laevis* tadpoles. Tadpoles from this flight that developed lordotic tails (see lower inset) revealed rVOR depression. G1g, 1 g-ground control (from [116])

In bull frogs, the electrical activity of the vestibular nerve fell shortly after onset of microgravity (OFO mission, 1970) but increased again slowly during the space flight to values beyond the pre-flight level [31]. Monkeys from the COSMOS program exhibited an increase in electrical sensitivity of central vestibular neurons during otolith stimulation by linear acceleration [52]. In toadfish, neurophysiological recordings after the end of a 16-day spaceflight showed an increased gain in activity in the eighth nerve in response to translatory stimulation of the animals [28]. Morphological studies on snail (*Biomphilaria glabrata*) otoconia as well as on synapse numbers in utricular hair cells of rats and central vestibular nuclei of swordtail fish (*Xiphophorus helleri*) supported the notion of vestibular sensitization as a basic mechanism of adaptation to microgravity (cf. [115, 116]).

The discovery of vestibular sensitization demonstrates the importance of comparative physiological studies during space flight for understanding how the human vestibular system adapts to space weightlessness. In fact, psychophysiological measurements in astronauts have shown that the tilt sensation during artificial gravity stimulation was more pronounced, *i.e.* sensitized, for two days after a spaceflight compared to pre-flight observations [49].

Age-related extent of adaptive processes. Some animal species revealed an age-related sensitivity of physiological responses to real and simulated microgravity. Examples include the vestibulo-ocular reflexes of young fish [227] and *Xenopus* tadpoles [114], or the postural reflexes of rat pups [201]. Following tail suspension, the mass of fast muscle in 9-month old rats recovered significantly faster during the period of reloading compared to that of 30-month old rats [122].

The most significant expression of age-related sensitivity to gravity deprivation is demonstrated by the "critical period", that is a period of life with a very high sensitivity to any sensory deprivation, while earlier and later periods of life are not affected (cf. [268]). Until now, critical periods of gravity sensory systems have been demonstrated in zebrafish [181] based on studies using RWV-clinorotation. Application of the same simulation technique to transgenic zebrafish expressing the gfp gene under the control of a beta-actin promoter revealed periods of high susceptibility to gene expression in a variety of organs such as the notochord, eye, somites, and rohon beard neurons [233] that might affect physiological adaptation during development. Recently, gravity-related critical periods were also described for vestibular and tail development in Xenopus based on spaceflight experiments [279].

Persistence of adaptive changes. After space flights, neuronal, sensory, skeletal and motor systems, as well as respiratory, cardiovascular and endocrine systems all underwent physiological modifications. However, the effects were never permanent; *within* a few days of return to 1 g-conditions, the initially observed modifications disappeared. A rhythmic burst-like activity of spinal motoneurons recorded from the ventral root of Xenopus tadpoles during fictive swimming returned to normal within 3–6 days [27]. In *Opsanus*, the microgravity-induced up-regulation of utricular afferents disappeared within 30 h of landing [28]. The position sensitive interneuron (PSI) of the gravity sensory system of crickets *(Acheta domesticus)* needed about 2 weeks for recovery [113]. Head-up swimming of swordtail fish *(Xiphophorus helleri)* disappeared within a few hours of the Shuttle landing. Following a 3 g-treatment, *Oreochromis* normalized its swimming behaviour within 3–5 days [206], and modified metabolic markers returned to normal within 5 days [236].

3.2.5.4 Fish as Models to Analyse Adaptation at the Gene Level

The European consortium ENFORM (European Network Using Fish as Osteoporosis Research Model) was formed with the aim of using small and rapidly developing species such as zebrafish (*Danio rerio*) or Medaka (*Oryzias latipes*) to study physiological adaptation to microgravity at the gene level [185, 186, 210]. The preferred whole-genome/whole-body approach benefits from the transparency of embryos and young fish. Pharmacological control studies with parathyroid hormone (PTH) and vitamin D had shown that the skeletons of these fish respond as expected: PTH treatment decreased bone formation, the PTH receptor was induced while arrestin ß2 was inhibited, and the PTH gene itself was down-regulated [1]. In contrast, vitamin D increased bone formation and down-regulated expression of the PTH receptor. Therefore, this model had a solid basis to study bone homeostasis and related gene expression in altered gravity.

Regulation of genes associated with bone formation and regulation of bone homeostasis was observed in *Oryzias* in both head and trunk after exposure to clinorotation or RPM [209]. Transgenic zebrafish lines carrying the gene for the fluorescent protein EGFP under the control of bone-specific gene regulatory regions were used for live-imaging of gene expression under a fluorescence microscope after 2–3 days of clinorotation (Fig. 3.34) [186].

Gene expression at the whole genome level using microarray technology revealed that after exposure of 5-day old larvae to clinorotation for 1 day, 113 genes were up – and 93 down-regulated. Some of these genes affected muscle and skeleton maintenance, the immune system and general metabolism. Genes coding for signalling molecules (Notch, Fgf) involved in skeletal, nervous and cardiovascular development were also modulated by RPM and RWV treatment. Exposure to

Fig. 3.34 Fluorescence of the *egr1*-GFP transgene in developing zebrafish embryos at 8 dpf in control (*left*, ventral view) and clinorotation-treated (*right*) larvae. *m* mandible; *r* retina; *h* hyoid; *pa* pharyngeal arches; *pf* pectoral fin

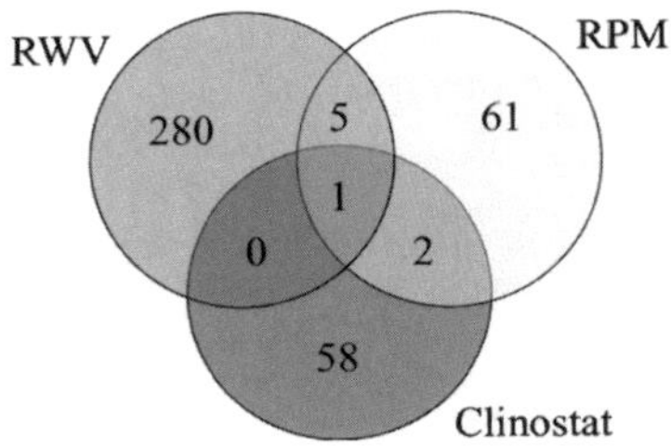

Fig. 3.35 Venn diagram representing numbers of genes with known function modulated in each of the microgravity simulation experiments (clinostat, RPM, RWV). Intersections represent the number of genes common to two out of three or all three experiments

3 g for 1 day by means of ESA's Large Diameter Centrifuge did not dramatically modify gene expression. Only a very small number of genes with a known function, either in humans, mice or rats, were obviously affected by two simulation conditions and only one by all three (Fig. 3.35). Special attention has to be paid to simulation techniques in order to avoid additional effects due to experimental conditions, as was also shown in a whole genome analysis of Drosophila in microgravity [106].

Due to improved technical resources (sequenced genome, large collection of mutants), ENFORM will be extending its whole-genome/whole-body approach to endothelial cell and vascular physiology [186]. The available gene expression data will give insights into physiological adaptation of the muscular, vestibular, vascular and immune systems. Other -omics (*e.g.* proteomics) approaches are envisaged as well as whole-body *in situ* hybridization or immunohistology. At the 5 dpf-stage now chosen, zebrafish larvae have hatched and are free-swimming, hunting and feeding mini-vertebrates that have most organs already in place and functioning. Many space-relevant physiological topics including microangiography, microdissection or labelling of individual neurons can be adapted to these fish.

ENFORM members are M. Muller, Liège/B, P. Aleström, Oslo/N, R. Goerlich, Aachen/D, C. Winkler, Würzburg/D and Singapore, K. Slenzka, OHB-Bremen/D and J. Maier, Milan/I.

3.2.5.5 Perspectives

The recent strategy in Space Life Science research with its dominant focus on molecular biology, genetics and radiation biology (cf. Fig. 3.32) is limiting because it neglects the fact that the stability of a body is maintained by interacting physiological mechanisms. For example, flight rats in 12 h light/12 h dark conditions evidenced a pronounced phase delay in body temperature, but not heart rate, compared to controls [82]. Vestibular function depends on feedback information from the cardiovascular, endocrine, sensory and motor systems. *Vice versa*, the vestibular system adjusts heart rate, blood pressure, immune responses and arousal [203].

It has to be questioned whether these interactions can be analysed exclusively by the study of genes relevant for the circadian or vestibular system. Molecular biology and genetics will significantly push forward the development of effective countermeasures. However, central integration of information originating in different organs is mandatory for the physiological stability of organisms. Thus, physiological tests are needed to evaluate the effectiveness of countermeasures. In fact, what we know about exercise and nutrition countermeasures designed to minimize muscle atrophy has come from physiological studies.

Physiological studies in animals face several obstacles. One is the fact that transport of animals into space needs survival systems that are adapted to the requirements of the different species. Another comes from animal welfare laws in Europe. They leave animal physiologists with mixed feelings that space agencies are reducing the number of physiological studies using animals to avoid conflicts with public opinion, because they are already under pressure to justify budgets for space activities. Organ cultures as *in-vitro* models with a close approximation to *in-vivo* conditions may (partially) overcome such problems and support physiological studies. In fact, isolated inner ears of mice were successfully exposed to simulated microgravity for 7 days because sensory organ morphology and robust marker protein expression in hair cells were preserved [100]. This success opens a way to study regenerative and adaptive physiological capacities in other organs and might, therefore, become a powerful tool for future physiological research in simulated or real microgravity, as well as hypergravity.

3.2.6 Human Physiology

Gilles Clément[1], Laurence Vico[2], Alexander Chouker[3]

[1]International Space University, Strasbourg, France
[2]University of Saint-Etienne, Saint-Etienne, France
[3]University of Munich, Munich, Germany

3.2.6.1 Introduction

Since the first human spaceflight exactly 50 years ago, the interest in human physiology in space has kept growing along with the space program and the opportunities it provided for flying more and more individuals on board capsules, shuttles, space stations, and soon suborbital space planes. Many discoveries on humans' capabilities to adapt to the extreme environment of space missions have been made, but our current knowledge on humans in space does not exceed stays of 14 continuous months, in one single individual, and the cumulative time in space of all space travellers, as of today, is comparable to the lifetime of one single individual [47]. The greatest test for human physiology will be the projected nearly 3 years round-trip to Mars, in isolated and confined habitats, with the crew experiencing several transitions in levels of gravity, dangerous radiation, and the challenges of landing and living on their own on another planet. The showstoppers of a Mars mission may be adverse reactions of the human body, such as decreased motor and sensory capabilities, bone loss, alteration in immune function mechanisms, or a depressed homeostatic level. European scientists, often in collaboration with scientists from other countries, are actively studying these adaptive changes and testing possible prevention means and countermeasures.

3.2.6.2 Sensorimotor System

Many questions about how humans adapt to weightlessness focus on the direct effect that weightlessness has on the sensorimotor system. In a 1 g Earth environment there are many sensory systems that can establish reference frames for coding information used in posture, movement and spatial orientation. These sensory modalities include visual cues from the environment, somatosensory cues from contact between the body and Earth-fixed objects, and gravitoinertial acceleration through both the otolith organs of the inner ear and abdominal viscera graviceptors. Not perceiving gravity has an obvious effect on otolith signals as well as on tactile and proprioceptive inputs. Furthermore, motor commands must significantly change for the same movement performed in μg compared to 1 g. The sensorimotor system must clearly adapt to these conditions in weightlessness.

There are other potential effects of the weightless environment, however, that are less related to the effects of gravity on the body itself and more to the context in which an individual must interpret sensory information and act. Brain functions have developed on an evolutionary time scale to deal with the specific constraints that gravity imposes on cognitive functions. For instance, the world in which we live is primarily two-dimensional (2D), particularly for Earth-bound creatures. While humans have constructed massive, three-dimensional (3D) structures such as skyscrapers, these edifices can essentially be described as multi-layered 2D environments. Neural processes that allow us to navigate within this world may thus be specialized for 2D spatial representations. On Earth, we also expect to see objects disposed in particular fashions within the environment: objects lying on a table will usually be found in a stable upright or horizontal position; free-falling objects accelerate downward; we usually meet people in an upright position. In building these expectations, we are essentially modelling the expected behaviour of objects in the world. These models can be used to predict upcoming events and optimize performance on a variety of cognitive tasks. Furthermore, the human central nervous system (CNS) may take advantage of invariants in a gravitational environment to provide efficient coupling between perception and action. These mechanisms could break down when applied in a novel gravitational environment [160].

Spatial disorientation, perceptual illusions, space motion sickness in orbit, as well as post-flight disorientation, nausea, and locomotion problems are the most overt neurovestibular problems of spaceflight. Research programs in this area are aimed at developing scientifically based countermeasures for these problems, which are generally most acute after transitions between gravitational environments. Unfortunately these problems always come at times when physical and cognitive performance is critical for safety and mission success. Post-flight symptoms have generally been more severe after six-month Mir and International Space Station (ISS) flights than after two-week Space Shuttle missions, demonstrating that some components of sensorimotor adaptation to μg take place over time scales of months, rather than weeks. Extravehicular activity and teleoperation of robotic arms and vehicles have become increasingly common during the construction of the ISS, and these also represent significant sensorimotor integration challenges. When considering plans for very long duration exploration missions to Mars, sensorimotor problems are anticipated whenever astronauts must make a sudden transition from μg to the 0.38 g Mars gravity and back, or from μg to an artificial gravity environment [196].

A total of 420 experiments on space neuroscience were identified between the flight of Vostok-3 in 1963 and the ISS Expedition-15 in 2007 [50]. About 60 of these experiments have been performed on board the ISS since permanent occupation began in October 2000 [69]. The first experiments on the sensorimotor system in orbit focused on the control of posture and the adaptive changes in antigravity muscle activity. Then, postural strategies in reaction to body and limb movements were investigated. Due to the complexity of modelling the effects of weightlessness on human balance, when both the sensors in the muscles, joints, and skin are altered, and the actuators themselves, the muscles, are disused or atrophied, the focus of research shifted to the more accessible study of compensatory eye movements [45, 46]. With the Space Shuttle and its required piloting skills, experiments on space motion sickness, spatial disorientation and eye-head coordination were then performed in orbit and immediately after landing. During the last decade, accompanying the advances in cognitive neuroscience research in ground-based laboratories, studies were performed on the Space Shuttle and Mir on mental rotation, face recognition, and vision science [152]. In fact, the first neuroscience experiment performed on the European Columbus module studied the effects of exposure to weightlessness on the perception of distance and depth, as well as the shape of 3D objects (Fig. 3.36). This study is still ongoing. Measurements are based on three primary tasks presented in a head-mounted display: (a) depth perception – subjects adjust the shape of geometric illusions or the depth of 3D objects with a finger trackball; (b) distance perception – subjects estimate and report absolute distances between objects in 3D photographic scenes; (c) and hand drawing – subjects write or draw objects and words with an electronic pen on a digitizing tablet, without visual feedback. Pre-, in-, and post-flight measurements have been obtained from crewmembers of Expeditions 17, 20, and 22. Preliminary results indicate that a 3D cube looks "normal" when its height is smaller and width larger than a normal cube; and that hand-drawn objects have greater height and smaller width than on the ground. Also, the asymmetry between vertical and horizontal distance perception seen on Earth, where vertical distance is overestimated, seems

Fig. 3.36 ESA Astronaut Paolo Nespoli performing the 3D-SPACE experiment on board the ISS. He interacts with 3D views presented in a head-mounted display (the hood is used to block any external light that could give orientation cues). The digitizing tablet is used for neuropsychological testing of writing and drawing geometrical figures in absence of visual feedback (Source: courtesy of ESA)

to disappear late in-flight. These results indicate that the absence of a gravitational reference in orbit is responsible for changes in the mental representation of physical space, particularly its height and depth dimensions [50].

The first experimental findings from orbit were small but eventually grew in complexity and completeness. For example, a switch from leg extensor muscle tone activity on Earth to leg flexor muscle tone activity in orbit, and a reversal of up-down asymmetries in eye movements were identified. These findings resulted in a tangle of elements developed in isolation, and the model was that each subsystem (the bandwidth of graviceptors, the functionality of the muscle stretch reflex, the control of the body's centre of mass, the potential conflicts among orientation cues) was subject to local optimization as opposed to fulfilling a global adaptation purpose. This was a typical bottom-up approach. The current view is that adaptation to weightlessness is more central, whereas there are no non-reversible peripheral changes. Adaptation to spaceflight would therefore typically rely on a top-down approach. In this model, the CNS is still composed of "black boxes", such as places where integration of sensory inputs takes place, and strict rules are followed before a percept or a decision for a given movement is made. Perception depends not only on the signal processing derived from the sensory organs, but also very largely on knowledge derived from past experience and a central representation of physical space. In the near future, in-orbit studies of brain mapping with surface electrodes and functional magnetic resonance imagery will eventually elucidate the cortical structures and the internal mechanisms that will realistically validate this model.

In fact, neuroscience experiments on board the ISS are already investigating the effect of weightlessness on spatial perception, spatial memory, and navigation at cortical level. Using the *Multi Electrodes Encephalogram Measurement Module* (MEEMM) of the European Physiology Module on board Columbus, neuroscientists are studying the brain activity that underlies cognitive processes involved in visuomotor tracking, perception of self-orientation, 3D navigation, and discrimination of the orientation of objects [42]. These tasks, combined with physiological and behavioural measures, assess changes in EEG signals, particularly in the prefrontal cortical area to identify perceptual re-organization during adaptation to the spaceflight conditions. The prefrontal cortex plays an important role in cognitive control and is vulnerable to stressors such as fatigue, sleep deprivation, or hypoxia. The results of these studies will help determine whether these conditions present a risk for astronauts during tasks requiring spatial orientation.

Valuable information regarding the re-adaptation of CNS function to Earth gravity following spaceflight also comes from ground-based neuroscience experiments comparing pre- and post-flight responses. For example, current experiments conducted in Star City and at NASA Johnson Space Center are assessing eye movements, motion perception, and the sense of verticality in returning crewmembers, to determine how the neural pathways between the otolith organs and the CNS have adapted to spaceflight conditions. Motion-based simulators coupled with virtual reality investigate the effect of exposure to weightlessness on an astronaut's perception of self-motion and tilt immediately after spaceflight, and evaluate how augmented reality might improve spatial awareness and performance.

Studies are also looking at a possible relationship between vestibular and cardiovascular deconditioning after spaceflight, because control of blood pressure is highly dependent on postural changes [272]. Ground-based studies on animals have also suggested a possible link between the lack of vestibular signals and bone demineralization. If correlation is found between post-flight otolith deconditioning and orthostatic intolerance or bone demineralization this would be another argument for providing an artificial gravity countermeasure during future long-duration exploration missions [48].

Space research continues to explore how the constraints of the Earth's gravitational environment have established the processes for the coupling between perception and action and the consequences of long-duration space missions on these vital functions. The development of new space experimental paradigms, combining behavioural, neuropsychological, electrophysiological and functional brain imagery investigations provides a new tool for testing the alteration in spatial cognition in various environments, as well as pathological conditions, and aging.

3.2.6.3 Skeletal System

Bone is an active tissue that undergoes continuous remodelling, including resorption of old bone followed by formation of new bone matrix. With this remodelling bone adapts to local mechanical demand – it becomes stronger when there is an increase in mechanical stress, such as during impact sports, and it is removed where there is a decrease in mechanical stress, such as in near weightlessness during orbital flight. Thus, bone adapts its structure to its function. X-ray images of bone using either dual absorptiometry (DXA) or peripheral quantitative computed tomography (pQCT) were compared before and after Russian, American and

European space missions. Following stays on board Mir lasting one to six months, pQCT analyses (DensiScan, Scanco Medical, Switzerland) indicated a rapid bone loss in weight-bearing bones, such as the tibia. This loss was higher in the trabecular than the cortical bone compartment. On the other hand, non weight-bearing bones, such as the radius, did not seem to be affected [258]. More recently, investigations were conducted in 14 astronauts who had spent four to six months on board the ISS [141]. In these crewmembers, bone mass densitometry (BMD) measured by DXA and bone density measured by QCT were found to have decreased in the proximal femur, in the same manner as reported after the Mir missions [191]. At the femoral neck, bone loss is mainly caused by a thinning of cortical bone. In the lumbar spine, total and trabecular BMD decreased by similar degrees, from 0.7% to 0.9% per month. It seems that there is a gradient in the bone mineral loss, which begins in the lumbar region and gradually becomes greater and greater in the lower limbs, including the femur and the tibia. Upper limbs are protected from bone loss, and an increased BMD was even reported for the skull [177]. Thus, the conditions of spaceflight induce a redistribution of mineral density within the skeleton (Fig. 3.37).

Bone formation and resorption activities can also be assessed by determination of bone markers in blood or urine. However, few astronauts have been tested with

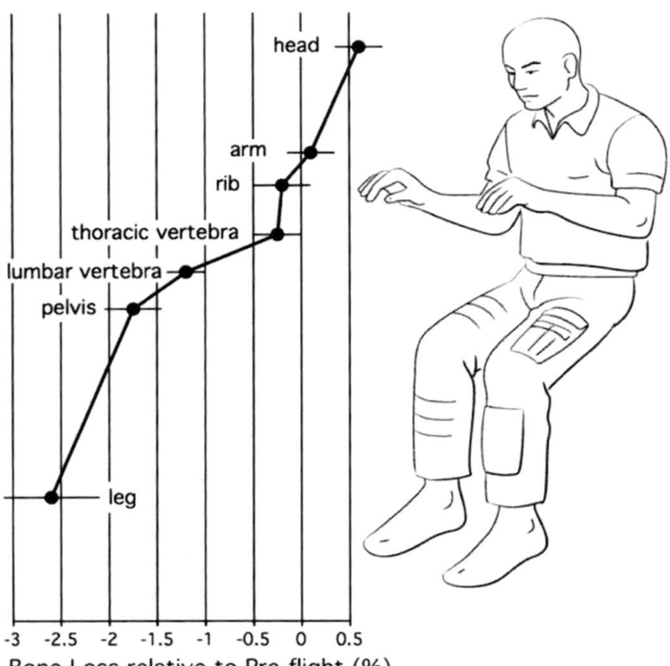

Fig. 3.37 Percentage of bone loss following spaceflights of at least 6-month duration compared to pre-flight baseline levels (Adapted from [47])

this method to date. Nevertheless, results suggest a significant uncoupling of bone turnover characterized by a sustained increase in bone resorption activity. Mild changes in hormones regulating bone metabolism appear to follow the alteration of bone cellular activities. Thus the skeleton seems to adapt to the new mechanical stress in weightlessness, but covariates, such as circadian rhythms, nutrition, and changes in other organs, may interfere at degrees that are still unknown. Ultimately, the problem with decreased bone mechanical properties is not due to the exposure to weightlessness, but to the return to Earth.

It is worth noting that individual BMD measurements revealed wide variability in bone alteration. However, no potential confounding factors have been found to account for this variability, including the cumulated time spent in microgravity for those astronauts who flew several missions. Targeted strategies designed to identify astronauts at particular risk of bone loss need to be developed, by taking into account the effects of environmental and lifestyle factors on genetic predisposition.

Recovery and re-adaptation capacities were analyzed at the distal radius and tibia by pQCT in 15 cosmonauts who had spent 2–6 months on board Mir during a survey period equivalent to the spaceflight duration [258]. The time needed to recover was found to be longer than the actual mission duration. Using data from 45 individual crewmembers, Sibonga et al. [234] developed a recovery model that predicts 50% restoration of bone loss for all sites within nine months after landing. Recovery analyzed at the proximal femur in 16 astronauts for 1 year after ISS missions lasting four to six months [142]. In addition to measurements of volumetric BMD, bone mineral content (BMC), bone volume, the sectional area of the femoral neck, and biomechanical indices were derived from these QCT measurements. After 1 year, the BMC of the proximal femur had normalized, but BMD and the bone mechanical properties had not completely recovered, which suggests a compensation of size and cross-sectional area. Altogether, these data show that the recovery time for bone integrity is much longer than the flight duration. It also seems that there is periosteal apposition compensation.

Space research on the skeletal system focuses on in-flight prevention of bone loss and long-term recovery. The process of re-adaptation itself may increase the risk of bone injury, as different skeletal components may recover at different rates. To diminish performance decrement and the risk of injury it is essential to develop integrated countermeasures that maintain muscle strength and endurance, bone mass, as well as postural balance. Efforts are being made on Earth to develop such countermeasures. For example, ESA's *Integrated Countermeasure and Rehabilitation Exerciser* (ICARE) includes an impact system that mimics in quality and quantity the daily impacts on bone on Earth. Generation of impacts is combined with eccentric contraction of muscles, which preserves the bone-muscle pair. This countermeasure allows resistive exercise with the possibility to control the resistance. The priority of this system is to train the lower part of the body. Drug strategies are also being tested. Research programs at NASA and JAXA are examining the effects of bisphosphonates, the most widely used anti-resorptive agents on Earth. Bisphosphonates are administered orally once a week during the flight (alendronate) or in a single-dose intravenous injection prior to the flight (zoledronate).

Crewmembers' bone structure is being measured before and after six-month stays on board the ISS. In the past 5 years tomography imaging has greatly evolved. A new version of the pQCT DensiScan (XtremeCT, Scanco Medical, Switzerland) has higher resolution allowing "virtual bone biopsies". This HR-pQCT has the potential to examine both cortical and trabecular volumetric BMD and micro-architecture non-invasively at the distal radius and tibia with a resolution of 100 μ. In addition to geometrical and micro-architectural parameters, the bio-mechanical consequences of the changes in bone can be assessed by segmented CT-scans, which build a 3D voxel-based micro finite element model.

Both short- and long-term post-flight period surveys are also planned during experiments involving European, French and German space agencies. Post-landing events might be particularly challenging since bone loss following limb immobilization has been described after bed rest in humans [213] and after spaceflight in rats [139]. For this reason, peripheral QCT, HR-pQCT, and bone marker analysis are performed within a tight timing schedule over a three-month period following landing. Then the follow-up continues on a more flexible schedule for each individual, up to 18 months after the mission, to appreciate how much time is required to recover the bone lost at the various sites and bone compartments. Because the recovery of bone mass may not necessarily translate to recovery of bone strength, the loading conditions are compared before, during, and after the space missions. The daily mechanical stress analysis is done using a portable accelerometer. The hypothesis is that those astronauts who exercise with higher impact loadings show accelerated bone strength recovery compared to the other astronauts who expose their bones to less mechanical stress.

3.2.6.4 Immunological System

The founder of modern stress research, Hans Selye, reported in his Nature article in 1936 that, if an organism is challenged by non-specific "nocuous agents" it will develop a typical syndrome "independent of the nature of the damaging agent" to which the organism is exposed. This stepwise and time-dependent stress-syndrome includes the immune system as "one observes a rapid decrease in size of the thymus, spleen, lymph glands" and "adrenals are greatly enlarged but regain their lipoid granules, while the medullary chromaffin cells show vacuolization (...) in favour of increased elaboration of adrenotropic principles" [229]. Selye described here for the first time the pathophysiological consequences of stress in biological systems thereby affecting two important and stress-sensitive organ entities: the immune and neuroendocrine systems. Since then more than 520 humans have been subjected to environmental challenges that Selye had likely never considered: spaceflight. Especially in the last decade, evidence has increased that weightless conditions can affect single immune cell functions, as well as global immune dysfunction through alteration of stress-sensitive auto-, para-, and neuroendocrine immune control mechanisms.

3.2.6.5 Isolated Cell Systems

T-lymphocytes are immune cells that are of crucial importance in the regulation of adaptive immune responses. Yet, their activation is reduced in weightlessness. The pathophysiologically relevant cellular pathways and their complex molecular mechanisms remain to be elucidated. Space investigations performed during the last decade indicate that weightlessness impairs communication between immune cells and results in altered intracellular signalling pathways. The distinct graviceptive signal transduction elements have been localized in different cell compartments: (a) on the cell surface (*e.g.* the interleukin-2 receptor regulating proliferation/differentiation of T-cells); (b) in the cytoplasm (*e.g.* protein kinase C enzyme family controlling function of other intracellular proteins); and (c) in the cell nucleus (*e.g.* genes regulating cellular processes including proliferation/differentiation). However, the primary molecular mechanisms of how weightlessness influences cell signalling remain subject to future research [252].

3.2.6.6 Animal Experiments

Observations made from cell cultures were extended to *in vivo* animal experiments, thereby allowing the analysis of cells taken from immune organs. In agreement with the observations on stress responses made by Selye 70 years ago, it was observed that mice flown on board a 13-day space mission had a reduced spleen and thymus mass, as well as a smaller number of splenic leukocytes compared to their ground controls. Moreover, the immune cells' responses when activated through T-cell receptor-dependent pathways were reduced, indicating immune suppressive mechanisms [9]. Interestingly, it was also shown from gene expression analyses that spaceflight resulted in significant changes in thymic mRNA expression of those genes that regulate stress, glucocorticoid receptor metabolism, and T-cell signalling activity. Collectively, the data in mammals show that T-cell distribution and function, as well as immune and stress-response relevant gene expression, are significantly modified in response to spaceflight [147] (Fig. 3.38).

3.2.6.7 Investigations in Humans

The consequences of multi-factorial effects of physical stressors were investigated in humans using blood, urine and saliva sampling analyses (Fig. 3.36). The emotional stress of spaceflight was monitored using questionnaires. A decrease in the functional activity of phagocytes, natural killer cells, T-lymphocytes, and their

Fig. 3.38 ESA Astronaut Frank De Winne removing a tray from the MELFI, an European-built –80°C freezer, to insert human biological samples as part of a nutritional status assessment experiment on board the ISS (Source: courtesy of ESA)

responsiveness to cell-signalling molecules (cytokines) was observed after spaceflight. Testing of peripheral leukocyte subsets, early T-cell activation and intracellular / secreted cytokine profiles indicated a reduction in the ratio of interferon-γ (a pro-inflammatory Th1-cytokine) to interleukin-10 (an anti-inflammatory Th2-cytokine). This Th1 $<$ Th2 shift is considered to mirror a Th2 weighted, anti-inflammatory immune response. Stress-permissive neuroendocrine responses were reflected by increased sympathetic activity to predominate after spaceflights lasting two weeks, while longer missions were characterized by glucocorticoid-mediated changes, including impaired circadianicity of cortisol secretion. Other stress-response systems such as the immunotropic endocannabinoid system (ECS) were shown to be regulated during both short- and long-duration space missions. Preliminary results from the ISS indicate an activation of the ECS during spaceflights of six months, which is paralleled lower immune activation markers. Moreover, ground-based confinement studies have confirmed the link between emotional stress and immune alterations [43, 55, 130, 183, 242].

Not only are the immune functions impaired during spaceflight, but pathogenic microbes – the immune system's "counterparts" – are also more virulent. Stress hormones that down-regulate immune responses in humans can also facilitate the growth of several bacterial strains [95]. The result is that the human immune homeostasis is hit twice and in different ways, by reducing immune performance on the one hand, and by enhancing the growth and virulence of pathogenic microorganisms on the other hand. Therefore, multi – and cross-disciplinary approaches are needed to complement immune-directed research, including microbiology, neurobiology, and radiation biology. The full scale of *in vitro* and animal experiment opportunities in space and on Earth should also be taken into account to test these interactions.

3.2.6.8 Integrative Physiology

Human physiology is composed of a totally integrated set of complex subsystems that maintain critical physiological parameters such as temperature, fluid balance, biological rhythms and electrolyte levels, at relative stable levels. Spaceflight experiments have demonstrated changes in these critical physiological parameters and processes. For example, body fluid shifts, changes in electrolyte balance, blood cell mass, hormone synthesis and hormone action have all been observed during spaceflight. Several complications could arise from these prolonged changes, from dehydration to abnormalities in cardiac function, such as heart rhythm irregularities, as well as the ability to respond to physical and emotional stress or to handle normal as well as emergency procedures *e.g.* after a Mars landing.

For example, recent studies have shown that food safety, food quality and nutrition play a major role in maintaining the crew's overall health and ensuring the success of an exploration mission [140, 241]. There is also extensive research literature supporting the role of nutritional factors in understanding the mechanisms of depression and other mental disorders [271, 273]. The composition, safety, quality and quantity of food and water may all influence the physical and mental health of crewmembers during space missions, particularly when these missions are of long duration.

It is critical to determine the optimal nutrient requirements for the crew of a space exploration mission, not only for the maintenance of their health during flight, but also after they return to Earth. However, knowledge of the nutrient requirements of human subjects in extra-terrestrial environments is currently limited to the data from the twelve astronauts who have walked on the Moon. Based on this information and recent space station data, it is likely that the nutritional intake in conjunction with the amount of physical activity undertaken by astronauts in μg and reduced gravity will affect their weight management, and plasma levels of lipids, vitamins and minerals, salts, protein and carbohydrate. Adequate dietary intake and synthesis of various nutrients involved in bone and calcium homeostasis, including calcium, vitamins D and K, protein, sodium and phosphorus are of fundamental importance [237]. In addition to an adequate quantity of food and water, the quality of nutrition may alter the homeostasis of red and white blood cells, as well as vascular integrity, bone mass, and osmolality [238]. The appearance and quality of food can also seriously affect crew morale [132].

It is now well established that astronauts' energy expenditure in space is unchanged or even increased compared to that on Earth. However, the energy intake of astronauts during space missions is 30–40% below their needs. This difference is not due to inadequate planning in terms of nutrient intake or unbalanced menus, but to constraints related to time for meal preparation and consumption, work overload, and lack of appetite [140]. In addition to insufficient energy intake, the level and metabolism of vitamin D were altered after long-duration missions, even when the vitamin was taken as a supplement during flight [238]. Altered vitamin D status, in turn, was associated with increased bone resorption after landing. Decreases of up to 45% in magnesium and phosphorus concentrations were also reported. The decrease in magnesium is a concern for long-duration missions because of its role in preventing formation of renal stones. Other vitamin levels are also affected by prolonged human exposure to the space environment. Vitamin K, an important element in maintaining bone health, was reported to decrease during spaceflight. Levels of vitamin E, which can act as anti-oxidant and signalling molecule, and folate also decreased during spaceflight [238]. This is an issue for long-duration missions because folate is the naturally occurring form of vitamin B, which is especially important in aiding rapid cell division and growth, and synthesizing and repairing DNA, and it may have a positive influence on cardiovascular disease, as well as on several types of cancer, dementia, and affective disorders.

Another concern is the reduction in red blood cell count after short – and long-term space missions, which can go as high as 15% after two weeks of flight [237]. This "space anemia" accompanies the overall decrease in body fluid volume, which also affects orthostatic tolerance during standing immediately after landing [36]. The levels of stored iron also increase during spaceflight. Although these levels return to normal within days of landing, increased levels of stored iron and availability in tissue would suggest a need to reduce iron intake during long-duration missions, given that the long-term effects of this condition are unknown.

Nutrition during space missions is not just a concern related to diet and its clinical aspects, but expands to the technology aspects of life support systems as well as human factors. In addition to the reduced body weight, there is a muscle loss as a result of reduced protein synthesis by inactive muscle [75]. The lack of use of the anti-gravity muscles for maintaining postural tone and for locomotion in μg is also responsible for reduced size and strength of the extensor muscle groups during spaceflight. It begins after a few days in orbit and continues at a slower rate throughout space missions [6]. Astronauts returning from missions as short as one or two weeks have exhibited lack of coordination, weakness and fatigue [64, 212]. These anatomical, structural and functional changes adversely influence performance on extended missions, and will pose a serious concern for exploration missions with surface activities. It is clear that countermeasures are required, including pharmacological, dietary, chemical and mechanical manipulations. Ground-based studies are currently testing whether artificial gravity generated by centrifugation during bed rest is an effective, efficient multi-system countermeasure to the risks associated with bone loss, cardiovascular deconditioning, muscle weakening, neuro-vestibular disturbances, space anemia and immune system deficiency [48].

3.2.6.9 Conclusion

To understand the adaptation of living organisms to the space environment, it is necessary to explore the evolutionary strategies of how biological entities cope with new challenges to life [247]. In particular, more integrative "holistic" experimental set-ups in humans will allow a deeper insight into the multi-factorial effects of spaceflight on human physiological functions. This knowledge will help to prepare humans for long-duration missions on the ISS and beyond, and will serve to better understand and treat nervous system, cardiovascular, musculoskeletal disorders, as well as nutrition, stress and immune system related diseases on Earth.

3.2.7 Psychology

Bob Hockey[1], Gro Mjeldheim Sandal[2]

[1]Department of Psychology, University of Sheffield, Norton, UK
[2]Department of Psychosocial Science, University of Bergen, Bergen, Norway

3.2.7.1 Introduction

Research on psychology has long been recognised as a major part of the need to understand the problems likely to be faced by manned space exploration. This has sometimes been explained by the belief that humans may be seen as the 'weak link' in the chain. This chapter reviews what is known about the vulnerability of humans to the problems of spaceflight, in terms of both performance and wellbeing.

3.2.7.2 The Psychological Environment of Space

Space missions are characterized by a number of distinctive physical stressors, some of which are real threats to health and survival (danger, radiation), and others less so (low gravity, noise and vibration, abnormal light-dark periodicity, monotony, extreme isolation and confinement, lack of privacy). There are also many other conditions that are similar to unwanted Earth experiences: highly demanding or unpredictable workloads, interpersonal difficulties with crew and ground staff, sleep disruption, and so on. While some of these problems fall within the domain of medicine or engineering, many require an analysis on the psychological level. This is because part of the impact of physical stressors is mediated through the individual's perception and appraisal of the environment; for example as a danger to life, or a threat to health and wellbeing, or to task performance [146, 245]. Such responses may have secondary impacts on performance and social interaction, through the generation of anxiety and fatigue states. This vulnerability of humans to the space environment has only been recognised in official space programmes in comparatively recent times [246], and may be contrasted with NASA's earlier model of the astronaut as "invincible superhero" ("the right stuff").

While there have been many reports of what appear to be genuine stress-related problems in actual missions [214, 231], mainly of disturbances of well-being and mood, there have, as yet, been no formal studies. Our main sources of evidence are from simulation studies based on analogue environments, such as research stations in Antarctica or confinement in hyperbaric chambers. While these are often able to mimic the main physical features of the environment (*e.g.*, isolation and confinement), they have the limitation that they may not be perceived as threatening. For example, Sandal [223] found that anxiety was much lower during simulations based on land-based hyperbaric chambers than for polar overwintering. Another limitation is that participants may not be comparable to astronauts in terms of professional training background, group size and other variables. Nevertheless, the large amount of research that has been conducted in laboratories and field studies indicates that there are very likely to be threats to mission goals from a failure to adapt to the space environment. Furthermore, since most of this evidence has been obtained under low threat conditions, it is highly probable that effects will be much greater under space conditions. The aspects of psychological functioning that have been examined most fully are those of routine perceptual and motor skills, cognitive functions associated with memory, thinking and decision making, and aspects of the crew's well-being and social environment. These are discussed in turn.

3.2.7.3 Human Performance

In practice, reports of major performance failings in operational tasks during space missions are very rare, although such problems may be underestimated because of a tendency for under-reporting. Unfortunately, few systematic studies have been conducted under actual space conditions, so that much of what we know comes from Earth-based simulations. The psychological processes underlying operational performance can be separated into two broad types: routinized perceptual-motor skills and complex cognitive functions.

Perceptual and Motor Skills

The most direct effect of the physical space environment is the observed difficulty caused by microgravity for the execution of sensorimotor processes. These include goal-directed head, limb or body movements; eye-hand coordination; high precision manipulation and movements; and the control of routine actions, such as those required for interacting with control panels or keyboards. Microgravity has been found to have a wide range of effects, including impairment of spatial orientation and the precision of movements. However, these are typically transient, and effectively nullified via adaptation through experience of working in the situation; see [128] for a review. In any case, such small-scale actions are normally impervious to disruption under any situation, since they are highly routinized as a result of their continuous use in everyday life.

More directly relevant to the success of mission goals are results indicating that complex perceptual-motor tasks may also be disrupted by exposure to space environments. These are activities such as compensatory tracking and two-hand coordination, where the input or target is changing position and requires guided movements under the control of attention [128]. There is also concern about forgetting important but little practiced skills, for example, during manual docking of spacecraft with Mir-23 [66, 219]. Several studies carried out on board Mir have shown decrements in these more complex skills during brief (up to 2 weeks) missions [165], though there is no data on longer periods of time spent in space. It is not clear at this stage how much of these more complex changes are attributable to the direct effects of microgravity, since they may be partially caused by factors such as sleep disturbances or generalized stress responses such as anxiety.

Cognitive Functions

It is likely that the greatest threat to human performance from the space environment would be to activities that are strongly dependent on cognitive functions (such as attention, memory, thinking and decision making). In contrast to routine perceptual motor skills, these depend for their effectiveness on the availability of a limited capacity for information management. In general, cognitive skills seem relatively unimpaired and intact during space missions, though analyses of crew errors during Mir missions have suggested a link between the occurrence of errors in mission tasks and changes in work-rest schedules, periods of high workload, and so on. Only a small number of studies have included cognitive tasks as a way of monitoring performance, all during short missions. Even when ground-based simulations are included [109, 110, 225], the currently available database is small. Again, there is little evidence of impairment, though some has been found when crewmembers have to carry out two tasks at the same time, or a single activity that makes heavy demands on executive processes [128]. Such effects are most evident during the first few weeks of long-duration space flight before adaptation occurs, suggesting that they are more likely to be stress-induced than a direct effect of microgravity or other aspects of the physical environment.

In Earth studies, cognitive impairments are common in high workload situations or under stress, though they may be masked by compensatory control activity, particularly when individuals are highly motivated or carrying out important work [108]. As a result, they can only be reliably detected through the use of a latent decrement methodology, in which secondary tasks or physiological measures are included to reveal the increased costs of protecting the primary task. Much further research is needed to examine the stability of cognitive performance during spaceflight. It will need to extend the period of exposure to several months, and, since such protection of task performance will be very strong, should also make use of methodologies based on the latent decrement approach.

3.2.7.4 Well-Being and Interpersonal Relationships

The success of most human space missions and the numerous examples of demanding tasks that have been carried out effectively are generally taken as evidence that astronauts can perform and cope in space, both as individuals and as teams. Yet, considerable anecdotal and behavioural evidence indicates that crewmembers have experienced psychological and interpersonal difficulties arising from the myriad stressors inherent in space missions, especially longer ones. Psychological and behavioural reactions have included emotional lability, psychosomatic symptoms, irritability toward crewmates and/or mission control staff, and a considerable decline in vigour and motivation [91, 128, 246]. Sleep disturbances and fatigue, as well as alterations of circadian rhythms in astronauts, are among the most important factors contributing to impaired well-being and alertness. Subjective reports from astronauts, as well as results from sleep monitoring studies, show that sleep in space is shorter, more disturbed, and often shallower than on Earth, though with a considerable degree of inter-individual variation [128].

A question of relevance to mission planning is whether the psychological resilience of crewmembers changes over the course of the mission. Only a few spaceflight studies have addressed this issue. However, results from long-duration missions suggest that sleeping problems occur primarily during the first 2–4 weeks as crew members adjust to the novelty of being in space. According to Russian experiences [91, 128], the most critical phase starts sometime between the 6th and 12th week of the mission, when the crewmembers settle into the routine of work in space. This corresponds to findings from analogue environments suggesting that the halfway point or the third quarter of the mission represent the most critical psychological phases. However, other studies suggest that the time course of adaptation depends on the kind of reactions studied, as well as on the specific features of the environmental conditions [223].

Astronaut crews represent small groups living close together in a remote and harsh habitat. The success and safety of space missions, as well as individual wellbeing, depend directly on the maintenance of crew cohesiveness, adequate leadership, and efficient co-working between crew members and between the crew and the Mission Control. The ability of crew members to communicate in an efficient manner is of critical importance to the accomplishment of mission goals and safety. Accident and incident investigations have demonstrated that communication failures can have devastating consequences [129]. A comprehensive study showed that communication problems due to misperceptions and misunderstandings, language problems, and differences in work style (*e.g.* decision making, preferred leadership) were among the most often mentioned challenges of 25 multinational space crews [250].

As mission duration and the complexity of operations increase and crews become more heterogeneous, psychosocial pressures are believed to take on added importance. Anecdotal evidence from space and simulation studies indicates that interpersonal tension is a correlate of confinement, frequently expressed indirectly; for example, through territorial behaviour, clique formation, or social withdrawal. While the potential deleterious impacts of interpersonal tension on crew performance have been recognized, less attention has been directed to the potential negative effects of too high a level of cohesion. Space crews often experience greater social cohesion over time by virtue of undergoing a common experience [128]. During long-duration missions involving high autonomy, crews may be vulnerable to the phenomenon known as "groupthink", characterized by symptoms involving illusions of invulnerability, reluctance to express concerns or disagreement about decisions, pressure towards conformity, and stereotyped views of people outside the group [224]. Groupthink has been associated with compromised performance and represents a safety risk. Aspects of groupthink have been observed during actual and simulated space missions. For example, crewmembers who do not adapt to group norms tend to become socially excluded from group activities, which again makes them prone to experience psychological health problems [128]. In addition, a tendency of crews to express aggression towards personnel in Mission Control has been interpreted as a mechanism for expressing negative feelings without disrupting their own relationships [224].

3.2.7.5 Conclusions

Despite the small number of studies carried out in actual space missions, there is considerable evidence that humans are mentally vulnerable to the demands of living and working in such an extreme environment. These problems need to be resolved since space exploration without humans is unlikely to be possible; they are the only part of the system capable of dynamic intelligence, and of context-based decision-making under unplanned or emergency circumstances. A number of ways of reducing these threats have been developed, including the screening and selection of astronauts, extensive training, and in-flight support of space crews. However, the demands of future missions, such as to Mars, will require a major investment in research on the understanding and management of both cognitive and social aspects of space psychology.

3.2.8 Research for Operational Space Medicine

Bernard Comet[1], Volker Damann[2], Audrey Berthier[1]

[1]Institute for Space Physiology and Medicine, BP 74404, 31405 Toulouse Cedex 4, France
[2]ESA/European Astronaut centre, Linder Hoehe, 51147 Cologne, Germany

3.2.8.1 Introduction

As described in previous sections, the human body adapts to weightlessness through a number of physiological changes, in particular those related to the bone-musculoskeletal system, the neuro-sensory system and the cardiovascular system. Some of the observed impairments, such as bone and muscle loss, are major issues still to be solved in view of future long-duration manned exploratory missions. In addition to this adaptation, the human body is subjected to new risks related to this specific environment, in particular radiation, as well as to the confinement situation, with risks of contamination, possible modification of the immune system and also a psychological impact on individuals and on group coherence and efficiency. Additionally, "common" health problems or diseases may also occur in-flight like on the ground.

The objective of Operational Space Medicine is to **maintain the health of the astronauts**. As defined by WHO, "Health is the physical, mental, and social well-being of humans". Operational Space Medicine involves crew surgeons (also called flight surgeons), who are employees of a Space Medicine Division in the space agencies. The mission of the crew surgeons is to optimize the health, fitness, and well-being of astronauts and their families, during their whole career with particular peaks of activity before, during and after the astronaut missions.

The following paragraphs describe the operational space medicine efforts to contribute to maintaining astronaut health, from prevention with use of counter-measures to diagnostic and care. Pre- and post-flight, as well as in-flight, standard physiological and medical exams and tests are conducted, as part of the usual medical follow-up of astronauts/cosmonauts.

3.2.8.2 Astronaut-Reported Medical Events

A historical and exhaustive view of astronaut-reported medical events (including flight and ground periods) is given in [14] and Sect. 3.2.2 and Chaps. 5 and 7. Twenty two astronauts/cosmonauts died while carrying out their job (18 during flight and four during ground simulation and testing). Three manned space missions have been interrupted for medical reasons (Salyut 5 in 1976, Salyut 7 in 1985 and MIR 2 in 1987). Excluding the "normal" human response observed during space flights (fluid shift observations, space adaptation syndrome, anthropometric changes, cardiovascular and volume changes, pulmonary function adaptation, bone and muscle changes, neurological, renal, gastrointestinal and endocrinal including clinical laboratory findings) that are explained in previous sections of this book, we discuss hereafter those which required in-flight medical care and which may warrant further research in the field.

The space adaptation syndrome (also called space motion sickness) is very common, 67% of astronauts are affected (among these affected astronauts, 19% severely, 35% moderately and 47% mildly). The medications used are Promethazine and Scopolamine which have significant side-effects.

Headache is also common during flight, accounting for complaints in 69% of flying male astronauts and 55% of flying female astronauts. Several factors might be involved including like fluid shift, CO_2 or CO levels. It is the most frequently given reason for consumption of pain relief medications.

Sleep disorders are also observed. Shift of timeline, stress, noise and other factors may be involved.... These disorders are controlled by non-benzodiazepine hypnotics and if necessary by benzodiazepine ones.

Skin problems have been reported including skin dryness in lower limbs, contact dermatitis, fungal infections and even one case of cellulitis. The available medication in the kits allowed the problem to be controlled [14].

In-flight, there is also an increase in the risk of eye injury and contamination. Several cases have been reported after elastic cord failure (scleral and corneal abrasions), or due to foreign bodies such as metal shavings when opening containers or devices. UV keratitis has also been reported when looking at the Earth without a filter. All these cases have been controlled with the available items in the medical kits [14]. It should also be noted that after flight, rare irreversible changes in vision have recently been reported and are under investigation.

Gastrointestinal complaints have been observed, such as frequent eructation, gastritis and oesophageal reflux probably due to changes in water – air separation in the stomach. Constipation is also a frequent complaint, probably caused by the absence of gravity reducing the progression speed of the digestive bolus. Infectious diarrheic episodes have rarely been reported. All these cases have been controlled with the available items in the medical kits [14].

Congestion of the nasopharyngeal sphere including the sinuses is an early complaint in space flight, due to the fluid shift. Viral infections of the upper respiratory system have been reported. All these cases have been controlled with the available items in the medical kits [14].

Some irritant pulmonary disorders due to fibreglass and exotic pollutants (smoke, ethylene glycol) have been observed. In 1975 during re-entry, accidental exposure to nitrogen tetroxide caused a pulmonary oedema with hospitalization after landing, but without long term sequelae. Masks with filter cartridges and necessary therapeutic means are available on board to control such emergency situations [14].

Pre-flight testing tries to identify for each crewmember if there is any allergic risk. Only local allergic skin reactions have been reported and have been controlled with the available items in the medical kits [14].

Urine retention (probably due to prostate adenoma and/or side effects of anti motion sickness medications), kidney stone episodes and prostatitis have also been reported (10 cases of kidney stones, one case of prostatitis [14] found after flight among 332 flown astronauts at NASA). One case of prostatitis with urine retention has a mission interruption. An additional intake of calcium has to be absolutely avoided as the bone is no longer fixing this additional calcium, it is excreted in the urine and increases strongly the risk of kidney stone formation. A Renal Stone Risk Assessment is now carried out for NASA astronauts.

Episodes of arrhythmia are often observed during space flights. An Apollo astronaut experienced during flight a 14 s bigeminy run and 2 years later he had a myocardial infarction. A Russian mission was interrupted for in-flight paroxysmal supraventricular tachycardia. Today on board the ISS the necessary means to achieve advanced cardiac life support (ACLS) are available and the crew medical officer is trained to use it [14].

Lumbar back pain is a common complaint among flight astronauts, often the cause of sudden awakening from sleep. This lumbar pain can be effectively controlled by body positioning during sleep (*i.e.* knee to chest position) and if necessary by use of pain relief medication.

Only superficial trauma (bruises, contusions) often during use of EVA suits (finger and nail bruises from EVA gloves) have required treatment. EVA can also be a source of muscular strain syndrome in the arms and shoulders. These strains and pains are well controlled by nonsteroidal anti-inflammatory agents and pain relief medication.

Superficial laceration is quickly cleaned and closed to prevent any further infection, a proposed estimate indicates that 50% of lacerations could be infected due to the level of micro-organisms in the spacecraft atmosphere.

Musculoskeletal traumas have been reported during EVA, during use of countermeasures or during unplanned troubleshooting activities. Ligament injuries (ankle, knee and wrist) have been reported often when there was a similar pre-flight history of injuries. Following such injuries, the in-flight use of countermeasures was very limited and had a serious impact on the post-flight health condition of the concerned subjects.

Although eight accidental events of on-board combustion have been reported (four on STS and four on Salyut and Mir) no cases of burns or smoke intoxication have been reported. Medical kits contain the necessary devices and medication to control such injuries for 24 h on one subject [14] (this will probably involve an emergency return).

3.2.8.3 Operational Health Prevention, Countermeasures and Care

"Prevention" and actions to counteract the impairments observed in weightlessness are implemented from astronaut selection, in pre-flight through specific training programs, in-flight through the use of on-board countermeasures and medical kits, and in post-flight to support recovery.

Astronaut selection includes both a medical and psychological selection phase with specific criteria and tests. It aims to avoid or minimize any additional risk due to health or psychological conditions that could worsen the health changes observed in space or that could be worsened by the flight environment and finally that could represent an additional risk to mission success.

Preparation, Training: Before Flight

Before flight, the preparation starts with a specific training program including basic medical training and mission-specific training. This program familiarizes astronauts with the anticipated space-induced health impairments and the means to counteract them. In addition, it familiarizes them with the most probable medical conditions that could occur in flight, with the methods to diagnose and treat them, and with the on-board medical kits and countermeasures. Astronauts have a yearly physical and mental check up. In addition, they have to maintain their physical fitness by regular physical exercise and sport during their whole career. When preparing for the missions, they also have access to physiological training to improve g tolerance (centrifuge training), space adaptation syndrome resistance (rotating chair training, parabolic flights, autogenic feedback training, training with tilt translation device, etc.), hypoxia tolerance (altitude chamber training) and familiarity with the extra vehicular space suits by training in a vacuum chamber and neutral buoyancy facility.

In-flight – Use of Countermeasures

In-flight, various means, so-called countermeasures, are used to counteract the impairments induced by weightlessness. Countermeasures include physical ones like physical training on specific devices, but also nutritional or pharmaceutical countermeasures. Many research activities are undertaken to develop more integrated countermeasures, such as artificial gravity, or to combine different types of countermeasures.

The operational use and evaluation of the first medical countermeasures started with the Salyut and Skylab space station programs (1973–1983). The countermeasures used during these flights mainly included protocols against fluid shift and orthostatic intolerance and devices for physical exercise. The "braslets", thigh cuffs, were used during the first days in microgravity, to prevent the discomfort effects of the microgravity-induced fluid shift, on the Russian side (Soyuz and Salyut). They are still used today by cosmonauts on the ISS. Other countermeasures to post-flight orthostatic intolerance included on-orbit LBNP (Low Body Negative Pressure) training before the return to Earth (Skylab LBNP, Salyut Chibis), the anti-g suit for re-entry on board STS and Soyuz ("karkasse" and later "quintavre" on the Russian side) and a fluid and salt load before re-entry. Devices for physical exercise included ergometers, treadmills and the Russian Penguin Suit (a constraint suit). From 1986 to 1999, on board the MIR station, most of the countermeasures used were the same as those used on board Salyut 6 and 7. The main improvements concerned the use profiles and procedures [37].

The countermeasures currently used on board the ISS are based on a 4-day cycle, involving 3 days of prescribed exercise and a fourth day for optional exercise or rest. On each exercise day, the crew exercises in two sessions of 1–1.5 h each for a total of 2–3 h of crew time per day (including equipment set-up and storage, clothing change, personal hygiene).

The available physical exercise devices on the ISS are mainly cycle ergometers, a treadmill and a resistive exercise device. The cycle ergometers include the American CEVIS (Cycle Ergometer with Vibration Isolation and Stabilization) and the Cycle ergometer "VB–3" (Russian). These ergometers are used as both aerobic and anaerobic exercise countermeasures. They can be used to train legs or arms, especially for maintaining lower body musculature or for EVA arm exercise training (Fig. 3.39).

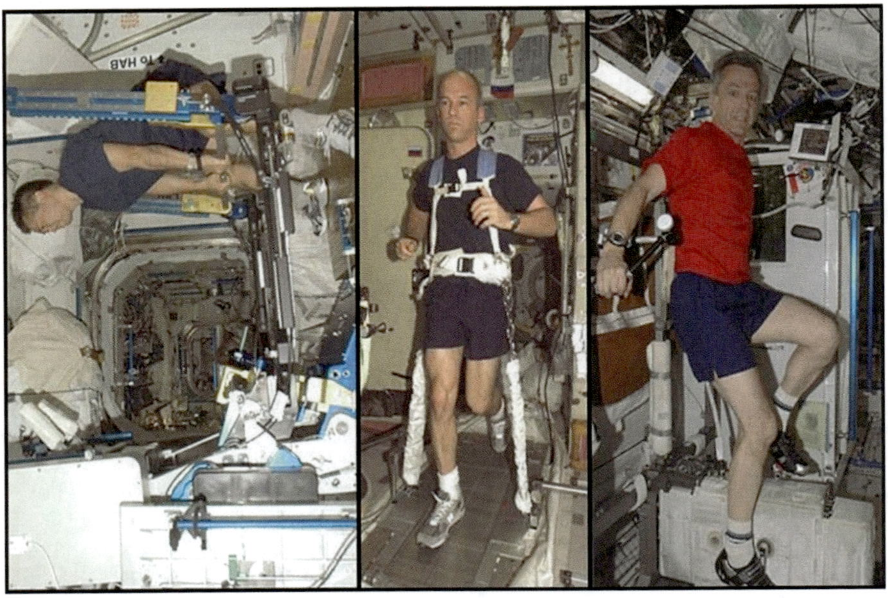

Fig. 3.39 The three exercisers used on the ISS, from *left* to *right* aRED, TVIS & CEVIS (Source: NASA)

The treadmill is the TVIS (Treadmill with Vibration Isolation and Stabilization), designed to allow walking and running, knee bends, and resistive exercise in a zero gravity environment. TVIS is used with a special harness called the Subject Loading Device (SLD) which provides the restraint system to secure the exercising subject to the treadmill. A new version "Treadmill-2" with a new SLD is under development by NASA.

Resistive training exercise is enabled by the Advanced Resistive Exercise Device (aRED), which provides higher levels of resistance-type training during flight than was possible with the previous countermeasures only. aRED enables eccentric and concentric contraction through a full range of motion from the following exercises: squats, dead-lifts, knee raises, hip abductions, hip adductions, leg curls, heel raises, bent over rows, upright rows, shoulder raises, shoulder presses, biceps curls, triceps extensions, wrist curls, and hammer throws.

"Braslet" devices, as described above, are still used by Russians to prevent the side-effects of the fluid shift at the beginning of a mission. Sets of bungee cords / Chest expanders / Mini-gym are also available. The Chibis Suit, as described above, is still used by Russians in pre-re-entry sessions to prevent post-landing orthostatic hypotension. "Penguin – 3" a constant-loading suit, as well as the "Tonus – 3" device – for myoelectrostimulation are also on board the ISS. During re-entry and landing, Shuttle anti-g suits and the Soyuz "Kentavr" are used in combination with a water and salt load just before re-entry.

Other Countermeasure devices have been evaluated in space through research or technical commissioning protocols but they are not, at the present time, used as "operational" countermeasures. These include the Transient Heel Loading Device (THLD) on MIR 1996 or the Flywheel on ISS 2007 etc.

Other devices are evaluated during ground simulations, such as VIBE or Galileo platforms (Vibrations), LBNP combined with the treadmill, Short Arm Centrifuge, Dual-track treadmill with virtual reality, Grasim (Gravity simulator) . . ..

Post-flight: Rehabilitation/Recovery

During the post flight period, health-care should allow a smooth return to the pre-flight health status. For recovery of cardiovascular and muscle capacities, physical exercise should be practiced, avoiding too strong and too rapid loading to prevent any back pain or muscular strains on the lower body muscles. For bone recovery, physical exercise with good mechanical constraints (running, brisk walking) is probably the most efficient way to achieve a full recovery.

3.2.8.4 Crew Health Maintenance Systems for Future Manned Space Exploratory Missions

Several studies and projects are contributing to develop new concepts to support crew health maintenance for future manned space exploratory missions. Besides the health issues raised by the environments (including radiation and psychological risks) during long exploratory missions, there is a significant probability of disease and injury. While for Moon missions, emergency return within a few days is possible, this will not be the case for Mars missions. For these missions, the crew will be faced with very severe confinement with no possibility of emergency return and very difficult conditions for remote assistance with a one-way communication delay of up to 20 min [118]. In that context, autonomy in preventing, diagnosing and treating any medical condition will be crucial. The crew will most likely include a Medical Doctor. But he/she should be supported by specific equipment to assist with on-board diagnosis, surgery and care. New research for instance on computer-aided diagnosis and surgery, on new diagnostic means, on medical robot support or on the use of virtual and augmented reality both to train and guide astronauts for medical acts should contribute to addressing this need for autonomy [166].

3.2.8.5 Current Research Needs for Operational Space Medicine

Considering the operational space medicine concerns, the following research issues can be listed and ranked in decreasing order of priority.

- Astronaut bone and muscle loss in-flight is still a topical medical issue and is already the subject of numerous research activities. The remaining questions are: What are the mechanisms? Is it reversible? Which optimized countermeasure should be used? And what post-flight optimized rehabilitation is needed?
- Control of infectious diseases (on-board micro-organism mutations, drug efficiency, drug resistance) should also be investigated: is it a critical issue for exploratory missions? Which solutions can be considered?
- Control of the acute risks of radiation exposure (solar flares). How to implement the ALAP "As Low As Possible" principle when designing extra-magnetosphere exploratory manned missions?
- In g-transitions & post-flight astronaut space motion sickness: which medication without side effects should be used?
- In-flight astronaut vision changes: is it a medical issue? What are the mechanisms? Is it reversible? Which prevention means should be used? What post-flight rehabilitation is needed?
- In-flight astronaut back pain is a medical issue: What are the mechanisms? What would be the optimized countermeasure?
- In-flight astronaut hearing changes: is it a medical issue? Is it reversible? What are the mechanisms? Which countermeasure should be considered?
- In-flight astronaut changes in CO_2 sensitivity: is it a medical issue? How can they be explained? Which countermeasure should be considered?
- Post-flight astronaut muscle atrophy, back pain, flat feet with plantar pain, leg weakness, spine muscular weakness: What post-flight optimized rehabilitation is needed?
- Post-flight astronaut cardiovascular reconditioning consequences: low exercise capacity of the lower limbs, oedema in the ankle – Is it a statistically valid medical issue? What would be the optimum post-flight rehabilitation?
- In-flight astronaut fluid shift: How to counteract facial and conjunctival oedema, the increase in transient intra-ocular pressure, nasal congestion, etc.. . .?
- In-flight astronaut transient increase in cerebral pressure: is it a medical issue?
- In-flight and post-flight astronaut laboratory-observed transitory changes, is it a medical issue? Why, and which prevention measures could be implemented?
- In-flight astronaut skin allergy and dryness: is it a medical issue? Which prevention measures should be used?
- In-flight astronaut changes in immune system: Is it a medical issue? What are the underlying mechanisms? Which prevention measures should be used?
- In-flight astronaut skin dryness of lower body parts (especially legs): is it a statistically valid medical observation? Why? Which in-flight prevention measures should be used?

In the future, long-duration missions will raise new challenges in terms of health-care during spaceflight. In parallel, new commercial spaceflights with space tourism will also raise new research questions, involving subjects who are not necessarily as healthy as astronauts/cosmonauts. Exploitation of existing data and future findings in the field should help answer these challenges.

3.2.9 *Research on Life Support Systems*

Joan Albiol[1], Cesare Lobascio[2], Maria Antonietta Perino[2]

[1]Universitat Autònoma de Barcelona, Barcelona, Spain
[2]Thales Alenia Space, Turin, Italy

Space environmental conditions are highly different from those found on the Earth's surface. Therefore, human space exploration requires the availability of a system that can sustain human life in hostile environments, commonly known as a life support system (LSS). The requirements for life support strongly depend on the mission's duration. To facilitate the description of the key issues it will be assumed that some of the requirements for human survivability, namely atmospheric pressure, environmental humidity, thermal control or radiation protection, are already provided by the confinement facility, whether this is a space transfer vehicle, space station or planetary base. Therefore they will not be further discussed. Likewise it will be assumed that there is an infrastructure in which the LSS will operate, providing all the required resources such as energy, data transmission and processing or available crew time.

A number of different technologies have been developed or are in development to fulfil LSS requirements. Most of those technologies rely on physico-chemical principles (physico-chemical life support systems, PhLSS). As will be discussed later, technologies based on biological systems (biological life support systems, BLSS) are becoming key elements for long-term space exploration missions. Nevertheless, the long-term LSS that will finally be used will most probably be a hybrid system combining the best characteristics of the two [57].

3.2.9.1 General Requirements to Sustain Human Life in Space

The objective of the LSS is to create a suitable environment to sustain human life, by controlling the environmental parameters, providing resources, and managing waste products [67]. Basic requirements on the LSS vary according to the mission scenarios and associated mission phases. These are affected by mission duration, crew size, availability of a source of resupply and feasibility or duration of rescue or evacuation actions. Typical atmospheric requirements include controlling humidity from 25% to 70%, CO_2 (PCO_2 0.705–1.011 kPa) and O_2 (19.5–23.1 kPa) levels. To meet those requirements the LSS will have to continuously provide resources and remove waste at rates proportional to the crew's metabolic activity. Table 3.2 – 1 lists the typical metabolic loads used for LSS design.

Table 3.2 1 typical metabolic loads used for LSS design [67]

Parameter	LSS loads	
	Standard value	Range
Crew O_2 consumption (kg/person/day)	0.84	0.49–1.25
Crew water consumption (kg/day/person)	2.8	Up to 5.15
Crew hygiene water usage (kg/day/person)	6.8	Up to 7.3
Crew urine production (kg/day/person)	1.56	Up to 2.0
Crew CO_2 generation rate (kg/person/day)	1.00	0.52–1.50

3.2.9.2 Life Support for Short-Term Extraterrestrial Missions

For short-term space missions, storage of resources or periodical resupply are feasible alternatives. Nevertheless, just considering the enormous cost of transporting mass far from the Earth's gravity field, recycling alternatives become the required choice, at least for the resources with the highest loads. Thus present day LSSs aim for different levels of water and atmospheric component regeneration. Due to technology readiness, safety and reliability considerations, regeneration today is mainly performed by means of physico-chemical systems.

3.2.9.3 Air Regeneration Subsystems

Oxygen is the most critical need in confined systems, where, in addition, it is necessary to control CO_2 and airborne contaminants to avoid toxicity for humans on board. On the ISS the total pressure is at atmospheric levels. Other spacecraft had lower total pressures for facilitating Extra Vehicular Activity operations. Oxygen is both re-supplied, *e.g.* via the European Automated Transfer Vehicle (ATV) and regenerated physico-chemically by means of Russian and US water electrolysers [267]. CO_2 is scrubbed via Four-Bed Molecular Sieve technology, and a Sabatier cycle has recently been implemented for recovery of O_2 from CO_2. On the ISS these functions are now centralized and provided by US and Russian partners, but advanced R&D on physico-chemical air revitalization is also ongoing in Europe and Japan. A complete European Air Revitalization System (ARES) is under development for possible addition on the ISS, including a solid amine CO_2 adsorber, water electrolyser and Sabatier equipment [24].

Trace gas contaminants are removed via a combination of charcoal and LiOH sorbent beds and a High Temperature Catalytic Oxidizer. For removal of airborne particulates and bio-contamination, High Efficiency Particulate Air (HEPA) filters are installed in the air loops in the different ISS modules.

Major constituents of the atmosphere are monitored by means of gas sensors (*e.g.* CO_2 in the Columbus module) and a centralized mass spectrometer, the Major Constituents Analyser, via a sampling system drawing air samples from the different ISS modules. A European FTIR analyser, the Analysing Interferometer for Ambient Air (ANITA) was operated successfully on the ISS from September 2007 to August 2008 [112].

3.2.9.4 Water Recovery Subsystems

Water is used by the crew for drinking, rehydrating food and hygiene purposes. Where implemented (*e.g.* on the ISS), it is also the basis for physico-chemical oxygen production via electrolysis. Several years of ISS experience have shown an average water balance of 3.65 kg per crew member per day [23], managed via resupply, storage and regeneration. Transfer of water from Earth, initially performed by the Russian Progress (Rodnik tanks) and the Shuttle via fuel-cell production, has seen the involvement of Europe become a reality with the ATV, able to transport up to 840 kg of potable water to the ISS, within three dedicated bladder tanks. Starting from water collected in Turin (Italy) and processed to meet quality and disinfection standards accepted by NASA and the RSA, Europe demonstrated its capability to successfully deliver this key resource to the ISS with the ATV Jules Verne mission [92]. With increasing levels of regeneration, and starting with condensate regeneration, the ISS water is processed [267] mainly via:

- Russian System for Water Recovery from Humidity Condensate (SRV-K) – multifiltration of condensate recovered from the atmosphere;
- Water Processor Assembly (WPA) – Multifiltration through ion-exchange resins and sorbent materials plus catalytic oxidation of trace organics;
- Urine Processor Assembly (UPA) – adopting Vapour Compression Distillation.

The Process Control & Water Quality Monitor checks the quality of the produced water for safe consumption, and dedicated storage systems (bellow tanks and flexible containers) are used [267].

3.2.9.5 Life Support for Long-Duration Manned Extraterrestrial Missions

Long-term manned exploration missions impose further constraints compared to short-term missions. Using present day technologies, any long-term mission would necessarily require a certain degree of resupply from Earth. For a Mars mission, not only the cost of resupply but also safety considerations impose self-sufficiency as a required condition. A 180-day reference Moon mission (four crew members) would require 767 kg O_2, 2,007 kg of food and 19,018 kg H_2O (including for hygiene and EVA), while a 1,000-day Mars mission scenario (six crew members, 525 days on Mars) would require 5,796 kg O_2, 15,171 kg of food and 147,496 kg H_2O [118]. In principle the enormous cost of supplying such an amount of mass from Earth can be heavily decreased with the use of a complete atmosphere and water recycling system, even taking into account the weight of the required equipment. The level of decrease depends on the technology used. The requirements would be even further reduced with *in situ* resource utilisation. For example, using water from the Moon or Mars surfaces, generating O_2 from the local water or growing higher plants on site. During transfer a major constraint would be the available space. But this is not foreseen for Mars or Moon surfaces. Present day physico-chemical technology does not allow for the design of a self-sufficient LSS system. Therefore major research efforts are being directed towards achieving complete recycling using both physico-chemical and biological alternatives.

As an example of additional physico-chemical water regeneration technologies, candidates for development to meet additional water needs on longer missions (*e.g.* a washing machine, a shower) include:

- Micro-, nano – and ultra–filtration processes
- Reverse Osmosis
- UV – Visible light photocatalysis
- Electrolytic ion separation (Flow-through Capacitor)
- Phase separation processes (besides distillation, freeze-drying)
- Electrolytic processes
- Biological processing of waste water and regeneration of gray water via higher plants (leaf transpiration)

3.2.9.6 Current Research in Bioregenerative Life Support

Although physico-chemical technologies are promising in terms of atmosphere revitalization and water regeneration, it would be extremely difficult to generate healthy food. On the other hand using self-regenerable biomass, such as higher plants, to generate food would not be only easier but at the same time would regenerate the atmosphere and provide most of the water. With the ideal goal of achieving complete recycling and self sufficiency of any manned outpost, all the different countries involved in space research are dedicating significant efforts to develop such bioregenerative life support systems.

A European Example in Bioregenerative Life Support

The foremost European effort in advanced life support is represented by the ESA MELiSSA (Microbial Ecological Life Support System Alternative) project. Initiated in 1988 [173] the MELiSSA system is a closed and controlled artificial ecosystem that was conceived as a tool to study and develop a Biological Life Support System (BLSS) for the future long-term manned extraterrestrial bases. Similarly to other approaches, the MELiSSA system relies on biological processes for food and water generation together with atmosphere revitalization. In its present configuration, a healthy crew diet is obtained using a higher plants compartment including wheat, tomato, potato, soybean, rice, spinach, onion or lettuce [266], complemented with the cyanobacteria *Arthrospira platensis* [143]. This compartment consumes the CO_2 produced by the crew while also regenerating O_2 and evapotranspirating H_2O which can be used as a source of clean water for the crew.

To fulfil complete closure, the MELiSSA system intends to use as far as possible biological compartments instead of physico-chemical alternatives. The baseline recycling components used in MELiSSA were inspired by the microbial communities found on a lake ecosystem [105]. This way the organic waste from the crew (faeces), together with the non-edible biomass from the photoautotrophic compartment and other biodegradable waste such as paper, are introduced into compartment I, also known as the thermophilic (55ºC) liquefying compartment. Here the organic waste is decomposed into more simple molecules such as acetate, CO_2, NH_3 and other related compounds. Some volatile fatty acids are also produced [104]. Minor recalcitrant biomass components, such as lignin derivatives, are degraded very slowly or not at all. For this reason, parallel complementary alternatives are also being investigated, either biological [155] or physico-chemical, such as a wet oxidation system.

The effluents from compartment I are further processed by compartment II containing anoxygenic photo-heterotrophic bacteria, where the major organic compounds are removed. Ammonium ion (NH_4^+), which can be used as a nitrogen source in any of the MELiSSA compartments but is not the preferred one for higher plants and *Arthrospira* growth, is converted into another form of nitrogen, nitrate (NO_3^-), using a consortium of nitrifying bacteria in compartment III [105]. The output of this compartment is a nutrient solution that can be directly used to provide mineral nutrients to higher plants and *Arthrospira* and therefore to close the liquid loop. Operational performance improvement and technology development is carried out in a dedicated facility known as the MELiSSA Pilot Plant [87, 88]. Intensive experimental tests, together with development of state-of-the-art monitoring and control systems [81] and advanced mathematical modelling [93] ensure successful implementation. The MELiSSA system has been proposed as an alternative BLSS for Moon [144] and Mars [249] planetary bases.

In parallel to the research performed on the MELiSSA system, a number of related aspects are being investigated within Europe. For example: alternatives for the first two compartments of the MELiSSA system, favouring the degradation of compounds into a gas phase instead of the liquid phase [38]; testing parts of the MELiSSA systems in space conditions [195]; how the build-up of small amounts of components in closed environments, *e.g.* due to inevitable microbial activities or to slowly leaking compounds from the construction materials, could be scrubbed by a biological air filter (Dutch BAF project) [207, 262]; or improving water quality such as in the Belgium BELLISSIMA, a scaled-down MELiSSA loop to study the evolution of small amounts of key contaminants among the compartments.

References

1. Aceto, J., Muller, M., Nourizadeh-Lillabadi, R., et al.: Small fish species as powerful model systems to study vertebrate physiology in space. J. Gravit. Physiol. **15**, P111–P112 (2008)
2. Albrecht-Buehler, G.: Possible mechanisms of indirect gravity sensing by cells. ASGSB Bull. **4**, 25–34 (1991)
3. ASGSB: Slideset. http://asgsb.org/slidesets/slidesets.html (2010)

4. Babbick, M., Dijkstra, C., Larkin, O.J., et al.: Expression of transcription factors after short-term exposure of Arabidopsis thaliana cell cultures to hypergravity and simulated microgravity (2-D/3-D clinorotation, magnetic levitation). Adv. Space Res. **39**, 1182–1189 (2007)
5. Balaban, E., Centini, C., Pompeiano, O.: Tonic gravity changes after gene expression in the efferent vestibular nucleus. Neuroreport **13**, 187–190 (2002)
6. Baldwin, K.M., Edgerton, V.R., Roy, R.R.: Muscle loss in space: physiological consequences. In: Mark, H., Salkin, M., Yousef, A. (eds.) Encyclopedia of Space Sciences and Technology. Wiley, Hoboken (2003)
7. Baluska, F., Kreibaum, A., Vitha, S., et al.: Central root cap cells are depleted of endoplasmic microtubules and actin microfilament bundles: implications for their role as gravity-sensing statocytes. Protoplasma **196**, 212–223 (1997)
8. Baluska, F., Mancuso, S., Volkmann, D., Barlow, P.W.: Root apex transition zone: a signalling-response nexus in the root. Trends Plant Sci. **15**, 402–408 (2010)
9. Baqai, F.P., Gridley, D.S., Slater, J.M., et al.: Effects of spaceflight on innate immune function and antioxidant gene expression. J. Appl. Physiol. **106**, 1935–1942 (2009)
10. Bar-Ilan, A., Rein, G., Fernandez-Pello, A.C., Torero, J.L., Urban, D.L.: Forced forward smoldering experiments in microgravity. Exp. Therm. Fluid Sci. **28**(7), 743–751 (2004)
11. Barjaktarović, Ž., Nordheim, A., Lamkemeyer, T., et al.: Time-course of changes in protein amounts of specific proteins upon exposure to hyper-g, 2-D clinorotation and random positioning of Arabidopsis thaliana cell cultures. J. Exp. Bot. **58**, 4357–4363 (2007)
12. Barjaktarović, Ž., Schütz, W., Madlung, J., et al.: Changes in the effective gravitational field strength affect the state of phosphorylation of stress-related proteins in callus cultures of Arabidopsis thaliana. J. Exp. Bot. **60**, 779–789 (2009)
13. Barmatz, M., Inseob, H.: Critical phenomena in microgravity: past, present, and future. Rev. Mod. Phys. **79**, 1–52 (2007)
14. Barratt, M.R., Pool Sam, L.: Principles of Clinical Medicine for Space Flight. Springer, New York (2008)
15. Benoit, M.R., Brown, R.B., Todd, P., et al.: Buoyant plumes from solute gradients generated by non-motile Escherichia coli. Phys. Biol. **5**, 046007 (2008)
16. Berry, M.V., Geim, A.K.: Of flying frogs and levitrons. Eur. J. Phys. **18**, 307–313 (1997)
17. Beysens, D., Chatain, D., Evesque, P., Garrabos, Y.: Beysens Maxus phases vib, High frequency driven capillary flows speed up the gas-liquid phase transition in zero-gravity conditions. Phys. Rev. Lett. **95**, 034502–034511 (2005)
18. Beysens, D., Evesque, P.: Vibrational phenomena in near-critical fluids and granular matter. In: Topical Teams in the Life & physical Sciences: Towards New Research Applications in Space, SP 1281. ESA, publication division, co ESTEC, Noordwijk (2005)
19. Billia, B., Fecht, H.J.: Microstructure and control in advanced casting processes. In: Seibert, G., et al. (eds.) A World Without Gravity – Research in Space for Health and Industrial Processes, pp. 186–210. European Space Agency, Noordwijk (2001)
20. Blakely, E.A., Kleiman, N.J., Neriishi, K., et al.: Radiation cataractogeensis: epidemiology and biology. Radiat. Res. **173**, 709–717 (2010)
21. Board on Physics and Astronomy: Connecting Quarks with the Cosmos: Eleven Science Questions for the New Century. The National Academies of Press, Washington, DC (2003)
22. Board on Physics and Astronomy: Revealing the Hidden Nature of Space and Time: Charting the Course for Elementary Particle Physics. The National Academies of Press, Washington, DC (2006)
23. Bobe, L.S., et al.: Performance of water recovery and water supplies system installed in ISS RS during ISS-1 – ISS-21 missions. AIAA ICES, Barcelona (2010)
24. Bockstahler. K., Funke, H., Brodt, K., Lucas, J.: Design Status of the Closed-Loop Air Revitalization System ARES for Accommodation on the ISS. AIAA ICES, Barcelona (2010)
25. Boonstra, J.: Growth factor-induced signal transduction in adherent mammalian cells is sensitive to gravity. FASEB J. **13**, S35–S42 (1999)

26. Borst, A.G., van Loon, J.J.W.A.: Technology and developments for the random positioning machine, RPM. Microgravity Sci. Technol. **21**, 287–292 (2009)
27. Böser, S., Dournon, C., Gualandris-Parisot, L., et al.: Altered gravity affects ventral root activity during fictive swimming and the static vestibuloocular reflex in young tadpoles (*Xenopus laevis*). Arch. Ital. Biol. **146**, 1–20 (2008)
28. Boyle, R., Mensinger, A.F., Yoshida, K., et al.: Neural readaptation to Earth's gravity following return from space. J. Neurophysiol. **86**, 2118–2122 (2000)
29. Boyle, R., Popova, Y., Varelas, J., et al.: Neurovestibular adaptation in the utricular otolith following extended periods of 3 g exposure and readaptation to 1 g. J. Gravit. Physiol. **15**, P37–P38 (2008)
30. Bozier, O., Veyssière, B.: Study of the mechanisms of dust suspension generation in a closed vessel under microgravity conditions. Microgravity Sci. Technol. **22**(2), 233–248 (2010)
31. Bracchi, F., Gualierotti, T., Morabito, A., et al.: Multiday recordings from the primary neurons of the statoreceptors of the labyrinth of the bullfrog. Acta Otolaryngol Stockh Suppl **334**, 3–27 (1975)
32. Brack, A., et al.: Do meteoroids of sedimentary origin survive terrestrial atmospheric entry? The ESA artificial meteorite experiment STONE. Planet. Space Sci. **50**, 763–772 (2002)
33. Bradamante, S., Villa, A., Versari, S., et al.: Oxidative stress and alterations in actin cytoskeleton trigger glutathione efflux in *Saccharomyces cerevisiae*. Biochim. Biophys. Acta **1803**, 1376–1385 (2010)
34. Brandstätter, F., et al.: Mineralogical alteration of artificial meteorites during atmospheric entry. The STONE-5 experiment. Planet. Space Sci. **56**, 976–984 (2008)
35. Briart, L.G., Maher, E.P.: Reserve utilization in seeds of Arabidopsis thaliana germinating in microgravity. Int. J. Plant Sci. **165**, 545–551 (2004)
36. Buckey Jr., J.C., Lane, L.D., Levine, B.D., et al.: Orthostatic intolerance after spaceflight. J. Appl. Physiol. **81**, 7–18 (1996)
37. Buckley, J.C.: Space Physiology. Oxford University Press, New York (2006)
38. Bursens, J., Verstraete, W., Albrecht, T., et al.: A Total Converting and Biosafe Liquefaction Compartment for MELiSSA. 35th International Conference on Environmental Systems. Rome, July 11-14. SAE paper 2005-01-3068 (http://papers.sae.org/2005-01-3068. DOI: 10.4271! /2005-01- (2005)
39. Castagnolo, D., Monti, R.: Thermal Marangoni flows. In: Monti, R. (ed) Earth Space Institute: Physics of Fluids in Microgravity. pp. 78–125. Taylor and Francis, London/New York (2001)
40. Chauveau, C., Birouk, M., Gökalp, I.: An analysis of the d2-law departure during droplet evaporation in microgravity. Int. J. of Multiphase Flow **37**(3), 252–259 (2011)
41. Chauveau, C., Gökalp, I., Segawa, D., Kadota, T., Enomoto, H.: Effects of reduced gravity on methanol droplet combustion at high pressures. Sym. (Int.) on Combust. **28**(1), 1071–1077 (2000)
42. Chéron, G., Leroya, A., De Saedeleera, C., et al.: Effect of gravity on human spontaneous 10-Hz electroencephalographic oscillations during the arrest reaction. Brain Res. **1121**, 104–116 (2006)
43. Choukèr, A., Kaufmann, I., Kreth, S., et al.: Motion sickness, stress and the endocannabinoid system. PLoS ONE **5**, e10752 (2010)
44. Chylack, L.T., Peterson, F.A.H., et al.: NASA study of cataract in astronauts (NASCA). Report 1: cross-sectional study of the relationship of exposure to space radiation and risk of lens opacity. Radiat. Res. **172**, 10–20 (2009)
45. Clarke, A.H.: Listing's plane and the otolith-mediated gravity vector. Prog. Brain Res. **171**, 291–294 (2008)
46. Clément, G.: Alteration of eye movements and motion perception in microgravity. Brain Res. Rev. **28**, 161–172 (1998)
47. Clément, G.: Fundamentals of Space Medicine, 2nd edn. Springer, New York (2011)
48. Clément, G., Bukley, A.P.: Artificial Gravity. Microcosm Press/Hawthorne/Springer, New York (2007)
49. Clément, G., Moore, S.T., Raphan, T., et al.: Perception of tilt (somatogravic illusion) in response to sustained linear acceleration during space flight. Exp. Brain Res. **138**, 410–418 (2001)

50. Clément, G., Reschke, M.F.: Neuroscience in Space. Springer, New York (2008)
51. Cockell, C.S.: The interplanetary exchange of photosynthesis. Orig. Life Evol. Biosph. **38**, 87–104 (2008)
52. Cohen, B., Yakushin, S.B., Holstein, G.R., et al.: Vestibular experiments in space. Adv. Space Biol. Med. **10**, 105–164 (2005)
53. Cottin, H., et al.: Heterogeneous solid/gas chemistry of organic compounds related to comets, meteorites, Titan and Mars: laboratory and in lower Earth orbit experiments. Adv. Space Res. **42**, 2019–2035 (2008)
54. Crabbe, A., Pycke, B., Van Houdt, R., et al.: Response of *Pseudomonas aeruginosa* PAO1 to low shear modelled microgravity involves AlgU regulation. Environ. Microbiol. **12**, 1545–1564 (2010)
55. Crucian, B.E., Stowe, R.P., Pierson, D.L., Sams, C.F.: Immune system dysregulation following short – vs long-duration spaceflight. Avia. Space Environ. Med. **79**, 835–843 (2008)
56. Cucinotta, F.A., Durante, M.: Cancer risk from exposure to galactic cosmic rays: implications for space exploration by human beings. Lancet Oncol. **7**, 431–435 (2006)
57. Czupalla, M., Horneck, G., Blome, H.J.: The conceptual design of a hybrid life support system based the evaluation and comparison of terrestrial test beds. Adv. Space Res. **35**, 1609–1620 (2005)
58. Daiker, V, Lebert, M. Häder, D-P., et al.: The involvement of a protein kinase in phototaxis and gravitaxis of Euglena gracilis. Planta. **233**, 1055–1062 (2011)
59. Demets, R., Schulte, W., Baglioni, P.: The past, present and future of BIOPAN. Adv. Space Res. **36**(2), 311–316 (2005)
60. Dreizin, E.L.: Experimental study of aluminum particle flame evolution in normal and micro-gravity. Combust. Flame **116**(3), 323–333 (1999)
61. Dupouy, M.D., Camel, D., Botalla, F., Abadie, J., Favier, J.J.: In: Thomas, B.G., Beckermann, C. (eds.) Proceedings of Modeling Welding and Advanced Solidification Processes, p. 415. TMS, Warrendale (1998)
62. Durante, M., Cucinotta, F.A.: Heavy ion carcinogenesis and human space exploration. Nat. Rev. Cancer **8**, 465–472 (2008)
63. Durante, M., Reitz, G., Angerer, O.: Space radiation research in Europe: flight experiments and ground-based studies. Radiat. Environ. Biophys. **49**, 295–302 (2010)
64. Edgerton, V.R., Roy, R.R.: Neuromuscular adaptation to actual and simulated weightlessness. Adv. Space Biol. Med. **4**, 33–67 (1994)
65. Ehrenfreund, P., Ruiterkamp, R., Peeters, Z., Foing, B., Salama, F., Martins, Z.: The ORGANICS experiments on BIOPAN V: UV and space exposure of aromatic compounds. Planet. Space Sci. **55**, 383 (2007)
66. Ellis, S.R.: Collision in space. Ergon Des. **8**(1), 4–9 (2000)
67. European Committee for Space Standardization: Environmental control and life support (ECLS). Space Engineering Standard, ECSS-E-ST-34C, http://www.ecss.nl/ (2008)
68. European Physics Society (EPS) position paper: The need for space flight opportunities in fundamental physics (PDF). EPS, http://www.eps.org/ (2005)
69. Evans, C.A., Robinson, J.A.: International space station science research accomplishments during the assembly years: an analysis of results from 2000-2008. NASA/TP-2009-213146-Revision A, NASA Johnson Space Center: Houston. Available in electronic form at http://ston.jsc.nasa.gov/collections/TRS/ (2009). Accessed 30 Nov 2010
70. Fasano, J.M., Massa, G.D., Gilroy, S.: Ionic signaling in plant responses to gravity and touch. J. Plant Growth Regul. **21**, 71–88 (2002)
71. Fecht, H.J.: Thermodynamic and mechanical properties of Zr-based bulk metallic glasses. Phil. Mag. B **76**(4), 495–503 (1997)
72. Fecht, H.J., Billia, B.: Breaking the Mould: Metallurgy in Microgravity, pp. 68–73. ASM American Society for metals, Cleveland, OH, New York, USA (2008)
73. Billia, B., Fecht, H.-J.: Microstructure and control in advanced casting processes. In: Fitton, B., Battrick, B. (eds.) A world without gravity - research in space for health and industrial processes. ESA SP-1251, European Space Agency, pp. 186–210 (2001)

74. Fecht, H.J., Billia, B.: Metals process simulation. In: Furrer, D., Semiatin, S.L. (eds.) Handbook ASM, vol. 22B, in print (2010)
75. Ferrando, A.A., Lane, H.W., Stuart, C.A., et al.: Prolonged bed rest decreases skeletal muscle and whole body protein synthesis. Am. J. Physiol. **270**, E627–E633 (1996)
76. Fortov, V.E., et al.: Complex (dusty) plasmas: current status, open issues, perspectives. Phys. Rep. **421**(1–2), 1–103 (2005)
77. Foucher, F., Fortov VE, Ivlev SA, Khrapak AG, and Morfill GE et al.: Testing the survival of microfossils in artificial martian sedimentary meteorites during entry into Earth's atmosphere: the STONE 6 experiment. Icarus **207**, 616–630 (2010)
78. Friedman, R. Gokoglu, S.A., Urba, D.L.: Microgravity combustion research: 1999 Program and results. NASA, Glenn Research Center, Cleveland, Ohio (1999)
79. Fuentes, A., Legros, G., Joulain, P., Vantelon, J.P., Torero, J.L.: Evaluation of the extinction factor in a laminar flame established over a PMMA plate in microgravity. Microgravity Sci. Technol. **17**(3), 10–14 (2005)
80. Fukuyama, H., Waseda, Y. (eds.): High-Temperature Measurements of Materials. Advances in Materials Research. Springer, Berlin (2009)
81. Fulget, N., Poughon, L., Richalet, J., et al.: MELISSA: global control strategy of the artificial ecosystem by using first principles models of the compartments. Adv. Space Res. **24**(3), 397–405 (1999)
82. Fuller, C.A.: How time flies - The effects of spaceflight on the circadian timing system. Gravit Space Biol **20**, 42 (2006)
83. Furrer, D.U., Fecht, H.J.: Nickel-base superalloys for turbine discs. J. Met. **1**, 14–17 (1999)
84. Gandin, Ch.-A. et. al.: Modeling of Heat and Solute Interactions upon Grain Structure Solidificaiton. Materials Science Forum. **649**, pp 189–198 (2010)
85. Garrabos, Y., Chabot, T.C., Wunenburger, R., Delville, J.P., Beysens, D.: Critical boiling phenomena observed in microgravity J. Chim. Phys. **96**, 1066–1073 (1999)
86. Gershuni, G.Z., Lyubimov, D.V.: Thermal Vibrational Convection. Wiley, New York (1998)
87. Godia, F., Albiol, J., Montesinos, J.L., et al.: MELISSA: a loop of interconnected bioreactors to develop life support in space. J. Biotechnol. **99**(3), 319–330 (2002)
88. Godia, F., Albiol, J., Perez, J., et al.: The MELISSA pilot plant facility as an integration test-bed for advanced life support systems. Adv. Space Res. **34**(7), 1483–1493 (2004)
89. Gorbunova, A.V.: Effects of space-flight factors on cytochemical characteristics of the motor analyzer neurons. Vestn. Ross. Akad. Med. Nauk **5**, 15–21 (2010). Russian
90. GoSpace – Industrial Research in Microgravity, The target of GoSpace as a one step integrated service is the stimulation of entrepreneurs for microgravity research. The initiative provides capabilities to analyze relevance of gravity to solve industrial research issues, give a user-friendly access for industrial researchers and investigate commercial market interest. Josef – Winner St. 1–3, D-58123 Bonn, See www.gospace.de
91. Grigoriev, A.I., Kozerenko, O.P. Myasnikov, V.I.: Selected problems of psychological support of prolonged space flights. Paper presented at the 38th congress of the international astronautical federation, Stockholm (1987)
92. Grizzaffi, L., Lobascio, C., Bruno, G., et al.: ATV water preparation campaign. 38th ICES, San Francisco (2008) SAE International. Paper is 'http://papers.sae.org/2008-01-2192
93. Gros, J.B., Poughon, L., Lasseur, C., et al.: Recycling efficiencies of C, H, O, N, S, and P elements in a biological life support system based on microorganisms and higher plants. Adv. Space Res. **31**(1), 195–199 (2003)
94. Guan, Y., Fray, N., Coll, P., Macari, F., Chaput, D., Raulin, F., Cottin, H.: UVolution: compared photochemistry of prebiotic organic compounds in low earth orbit and in the laboratory. Planet. Space Sci. **58**, 1327–1346 (2010)
95. Guéguinou, N., Huin-Schohn, C., Bascove, M., et al.: Could spaceflight-associated immune system weakening preclude the expansion of human presence beyond Earth's orbit? J. Leukoc. Biol. **86**, 1027–1038 (2009)
96. Gundlin, R.: Private communication. ACCESS Aachen, Germany. Final report EU/ESA integrated project IMPRESS (2008)

97. Günther, B.: Final report EU/ESA integrated project IMPRESS (2008)
98. Güttler, C., Krause, M., Geretshauser, R.J., Speith, R., Blum, J.: The physics of protoplanetesimal dust agglomerates. IV. Towards a dynamical collision model. Astrophys. J. **701**, 130–141 (2009)
99. Häder, D.P., Hemmersbach, R., Lebert, M.: Gravity and the Behaviour of Unicellular Organisms. University Press, Cambridge (2005)
100. Hahn, H., Müller, M., Löwenheim, H.: Whole organ culture of the postnatal sensory inner ear in simulated microgravity. J. Neurosci. Meth. **171**, 60–71 (2008)
101. Hampp, R., Hoffmann, E., Schönherr, K., et al.: Fusion and metabolism of plant cells as affected by microgravity. Planta **203**, S42–S53 (1997)
102. Haupt, A., Straub, J.: Evaluation of the isochoric heat capacity measurements at the critical isochore of SF_6 performed during the German Spacelab Mission D-2. Phys. Rev. E **59**, 1795–1802 (1999)
103. Hemmersbach, R., Braun, M.: Gravity-sensing and gravity-related signalling pathways in unicellular model systems of protists and plants. Signal Transduct **6**, 432–442 (2006)
104. Hendrickx, L., Mergeay, M.: From the deep sea to the stars: human life support through minimal communities current opinion. Microbiology **10**, 231–237 (2007)
105. Hendrickx, L., De Wever, H., Hermans, V., et al.: Microbial ecology of the closed artificial ecosystem MELiSSA (Micro-Ecological Life Support System Alternative): reinventing and compartmentalizing the earth's food and oxygen regeneration system for long-haul space exploration missions. Res. Microbiol. **157**, 77–86 (2006)
106. Herranz, R., Benguria, A., et al.: Spaceflight-related suboptimal conditions can accentuate the altered gravity response of Drosophila transcriptome. Mol. Ecol. **19**, 4255–4264 (2010)
107. Hinrichsen, H., Wolf, D.E. (eds.): The Physics of Granular Media. Wiley-VCH Verlag GmbH & Co, Weinheim (2004). ISBN 978-3-527-60362-6
108. Hockey, G.R.J.: Compensatory control in the regulation of human performance under stress and high workload: a cognitive energetical framework. Biol. Psychol. **45**, 73–93 (1997)
109. Hockey, G.R.J., Sauer, J.: Cognitive fatigue and complex decision-making under prolonged isolation and confinement. In: Bonting, S.L. (ed.) Advances in Space Biology and Medicine, vol. 5. JAI Press, Greenwich (1996)
110. Hockey, G.R.J., Wiethoff, M.: Cognitive fatigue in complex decision making. In: Bonting, S.L. (ed.) Advances in Space Biology and Medicine, vol. 3. JAI Press, Greenwich (1993)
111. Hofmeister, P., Blum, J., Heißelmann, D., The flow of granular matter under reduced-gravity conditions. In: Powders & Grains (2009)
112. Honne, A., Schumann-Olsen, H., Kaspersen, et al.: ANITA Air monitoring on the international space station part 2: air analyses. SAE Int. J. Aerosp **1**, 178–192 (2009)
113. Horn, E.R.: The development of gravity sensory systems during periods of altered gravity dependent sensory input. Adv. Space Biol. Med. **9**, 133–171 (2003)
114. Horn, E.: Microgravity-induced modifications of the vestibuloocular reflex in *Xenopus laevis* tadpoles are related to development and the occurrence of tail lordosis. J. Exp. Biol. **209**, 2847–2858 (2006)
115. Horn, E.: Signal transduction in vestibular adaptation to microgravity - A still unsolved problem. Signal Transduct. **7**, 240–247 (2007)
116. Horn, E.: Sensitization as a basic principle of vestibular adaptation to microgravity. J. Gravit. Physiol. **15**, P27–P28 (2008)
117. Horn, E., Böser, S., Franz, M., et al.: Development of the flight hardware for the experiment XENOPUS on the Kubik BIO4-mission. Microgravity Sci. Technol. (2010). doi:10.1007/s12217-010-9182
118. Horneck, G., Facius, R., Reichert, M., et al.: HUMEX, A Study on the Survivability and Adaptation of Humans to Long-Duration Exploratory Missions, ESA SP 1264. ESA-ESTEC, Noordwijk (2003)

119. Horneck, G., Klaus, D., Mancinelli, R.: Space microbiology. Microbiol. Mol. Rev. **2010**, 121–156 (2010)
120. Hu, X., Neill, S.J., Tang, Z., Cai, W.: Nitric oxide mediates gravitropic bending in soy-bean roots. Plant Physiol. **137**, 663–670 (2005)
121. Hughes-Fulford, M., Boonstra, J.: Cell mechanotransduction: cytoskeleton and related signalling pathways. In: Monici, M., Van Loon, J. (eds.) Cell Mechanochemistry, Biological Systems and Factors Inducing Mechanical Stress, pp. 75–95. Transword Research Network, Trivandrum (2010). ISBN 978-81-7895-458-5
122. Hwee, D.T., Bodine, S.C.: Age-related deficit in load-induced skeletal muscle growth. J. Gerontol. Biol. Sci. Med. Sci. **64**, 618–628 (2009)
123. Iida, T., Guthrie, R.I.L.: The Physical Properties of Liquid Metals, pp. 168–169. Clarendon Press, Oxford (1988)
124. Jiang, J.: MD – simulation of the atomic structure of glassy Zr-Cu metals. Appl. Phys. Lett. **92**, 011902 (2008)
125. Joo, H.J., Bae, Y.S., Lee, J.S.: Role of auxin-induced reactive oxygen species in root gravitropism. Plant Physiol. **126**, 1055–1060 (2001)
126. Joulain, P., Vietoris, T., Torero, J.: Gas-gas and gas-solid laminar flat plate diffusion flame in microgravity: Structure and stability. Microgravity Sci. Technol. **13**(1), 3–7 (2001)
127. Kamada, M., Higashitani, A., Ishioka, N.: Proteomic analysis of Arabidopsis root gravitropism. Biol. Sci. Space **19**, 148–154 (2005)
128. Kanas, N., Manzey, D.: Space Psychology and Psychiatry, 2nd edn. Kluwer, Dordrecht (2008)
129. Kanki, B., Rogers, D., Bessone, L., Parke, B., Sandal, G.M., Whiteley, I.: Team performance and space safety. J Br Interpl Soc **62**(7/8), 273–281 (2009)
130. Kaufmann, I., Schachtner, T., Feuerecker, M., et al.: Parabolic flight primes cytotoxic capabilities of polymorphonuclear leucocytes in humans. Eur J Clin Invest **39**, 723–728 (2009)
131. Kelton, K.F.: Crystal nucleation in liquids and glasses. In: Ehrenreich, H., Turnbull, D. (eds.), Solid state physics. Vol. **45**, pp. 75–179. Academic Press, New York (1991)
132. Kerwin, J., Seddon, R.: Eating in space - from an astronaut perspective. Nutrition **18**, 921–925 (2002)
133. Klaus, D., Simske, S., Todd, P., et al.: Investigation of space flight effects on Escherichia coli and a proposed model of underlying physical mechanisms. Microbiology **143**, 449–455 (1997)
134. Klaus, D.M.: Clinostats and bioreactors. Gravit. Space Biol. Bull. **14**, 55–64 (2001)
135. Krause, M., Bräucker, R., Hemmersbach, R.: Gravikinesis in Stylonychia mytilus is based on membrane potential changes. J. Exp. Biol. **213**, 161–171 (2010)
136. Kuang, A., Popova, A., McClure, G., Musgrave, M.E.: Dynamics of storage reserve deposition during Brassica rapa L. pollen and seed development in microgravity. Int. J. Plant Sci. **166**, 85–96 (2005)
137. Kumagai, S.: Combustion of fuel droplets in a falling chamber with special reference to the effect of natural convection. Jet Propulsion **26**(9), 786–786 (1956)
138. Kumar, G., Thang, H.X., Schroers, J.: Nanomoulding with amorphous metals. Nature **457**, 868 (2009)
139. Lafage-Proust, M.H., Collet, P., Dubost, J.M., et al.: Space-related bone mineral redistribution and lack of bone mass recovery after reambulation in young rats. Am. J. Physiol. **274**, R324–R334 (1998)
140. Lane, H.W., Feeback, D.L.: History of nutrition in space flight: overview. Nutrition **18**, 797–804 (2002)
141. Lang, T., LeBlanc, A., Evans, H., et al.: Cortical and trabecular bone mineral loss from the spine and hip in long-duration spaceflight. J. Bone Miner. Res. **19**, 1006–1012 (2004)
142. Lang, T.F., Leblanc, A.D., Evans, H.J., et al.: Adaptation of the proximal femur to skeletal reloading after long-duration spaceflight. J. Bone Miner. Res. **21**, 1224–1130 (2006)

143. Lasseur, C., Brunet, J., de Weever, H., et al.: MELiSSA: the European project of closed life support system. Gravit. Space Biol. **23**(2), 3–12 (2010)
144. Lasseur, C., Verstraete, W., Gros, J.B., et al.: MELISSA: a potential experiment for a precursor mission to the Moon. Adv. Space Res. **18**(11), 111–117 (1996)
145. Law, C.K., Faeth, G.M.: Opportunities and challenges of combustion in microgravity. Prog. Energy Combust. Sci. **20**(1), 65–113 (1994)
146. Lazarus, R.S.: Emotion and Adaptation. Oxford Univ. Press, New York (1991)
147. Lebsack, T.W., Fa, V., Woods, C.C., et al.: Microarray analysis of spaceflown murine thymus tissue reveals changes in gene expression regulating stress and glucocorticoid receptors. Cell Biochem **110**, 372–381 (2010)
148. Lebaigue, O., Colin, C., Larue de Tournemine, A.: Forced convection boiling and condensation of ammonia in microgravity **13**, 250–259 (2006)
149. Legrand, B., Chauveau, C., Shafirovich, E., Goldshleger, U., Carrea, E., Mounaim-Rousselle, C., Rouan, J.P., Gökalp, I.: Combustion of magnesium particles in carbon dioxide under microgravity conditions. J. Phys. IV **11**(6), 311–314 (2001)
150. Legros, G., Joulain, P., Vantelon, J.P., Fuentes, A., Bertheau, D., Torero, J.: Soot volume fraction measurements in a three-dimensional laminar diffusion flame established in microgravity. Combustion Science and Technology **178**(5), 813–835 (2006)
151. Leitz, G., Kang, B.-H., Schoenwaelder, M.E.A., Staehelin, L.A.: Statolith sedimentation ki-netics and force transduction to the cortical endoplasmic reticulum in gravity-sensing Ara-bidopsis columella cells. Plant Cell **21**, 843–860 (2009)
152. Léone, G.: The effect of gravity on human recognition of disoriented objects. Brain Res. Rev. **28**, 203–214 (1998)
153. Leroy, B., Rosier, C., Erculisse, V., et al.: Differential proteomic analysis using isotope-coded protein-labeling strategies: comparison, improvements and application to simulated microgravity effect on Cupriavidus metallidurans CH34. Proteomics **10**, 2281–2291 (2010)
154. Leys, N., Baatout, S., Rosier, C., et al.: The response of Cupriavidus metallidurans CH34 to spaceflight in the international space station. Antonie Leeuwenhoek **96**, 227–245 (2009)
155. Lissens, G., Verstraete, W., Albrecht, T., et al.: Advanced Anaerobic Bioconversion of Lignocellulosic Waste for Bioregenerative Life Support Following Thermal Water Treatment and Biodegradation by Fibrobacter Succinogenes. Biodegradation **15**(3), 173–183 (2004)
156. van Loon, J.J.W.A., Van Laar, M.C., Korterik, J.P., et al.: An atomic force microscope operating at hypergravity for in situ measurement of cellular mechano-response. J. Microsc. **233**, 234–243 (2009)
157. van Loon, J.J.W.A.: Some history and use of the Random Positioning Machine, RPM, in gravity related research. Adv. Space Res. **39**, 1161–1165 (2007)
158. van Loon, J.J.W.A.: The gravity environment in space experiments. In: Brinckmann, E. (ed.) Biology in Space and Life on Earth, pp. 17–32. Wiley-VCH, Weinheim (2007)
159. van Loon, J.J.W.A.: Mechanomics and physicomics in gravisensing. Microgravity Sci. Technol. **21**, 159–167 (2009)
160. McIntyre, J., Zago, M., Berthoz, A., et al.: Does the brain model Newton's laws? Nat. Neurosci. **4**, 693–695 (2001)
161. Mader, M., Misbah, C., Podgorski, T.: Dynamics and rheology of vesicles under a shear flow under gravity and microgravity. Microgravity Sci. Technol. **XVIII**(3/4), 200–2003 (2006)
162. Mancuso, S., Barlow, P.W., Volkmann, D., Baluska, F.: Actin turnover-mediated gravity response in maize root apices: gravitropism of decapped roots implicates gravisensing outside of the root cap. Plant Signal Behav. **1**, 52–58 (2006)
163. Mancuso, S., Boselli, M.: Characterisation of the oxygen fluxes in the division, elongation and mature zones of Vitis roots: influence of oxygen availability. Planta **214**, 767–774 (2002)
164. Manzano, A.I., et al.: Germination of Arabidopsis Seed in space and in simulated microgravity: alterations in root cell growth and proliferation. Microgravity Sci. Technol. **21**, 293–207 (2009)

165. Manzey, D.: Monitoring of mental performance during spaceflight. Aviat Space Env Med **71**, A69–A75 (2000)
166. Marstechcare Final Report: ESA Contract ESTEC N°: 16423/02/NL/LvH (2002)
167. Martzivanou, M., Hampp, R.: Hyper-gravity effects on the Arabidopsis transcriptome. Physiol. Plant. **118**, 221–231 (2003)
168. Matía, I., González-Camacho, F., Herranz, R., et al.: Plant cell proliferation and growth are altered by microgravity conditions in spaceflight. J. Plant Physiol. **167**, 184–193 (2010)
169. Mariotti, M., Maier, J.A.: Gravitational unloading induces an anti-angiogenic phenotype in human microvascular endothelial cells. J. Cell. Biochem. **104**, 129–135 (2008)
170. Martinez, I., Perales, J.: Mechanical Behavior of liquid bridges in microgravity. In: Monti, R (ed) Earth Space Institute: Physics of Fluids in Microgravity. pp. 21–45. Taylor and Francis, London/New York (2001)
171. Mastroleo, F., Van Houdt, R., Leroy, B., et al.: Experimental design and environmental parameters affect Rhodospirillum rubrum S1H response to space flight. ISME J. **3**, 1402–1419 (2009)
172. Medina, F.J., Herranz, R.: Microgravity environment uncouples cell growth and cell proliferation in root meristematic cells: the mediator role of auxin. Plant Signal Behav. **5**, 176–179 (2010)
173. Mergeay, M., Verstraete, W., Dubertret, G., et al.: MELISSA – A micro-organisms based model for CELSS development. In: Proceedings of the 3 rd European Symposium on Space Thermal Control & Life Support System, Noordwijk, European Space Agency!' s -288. (1988)
174. Mialdun, A., Ryzhkov, I.I., Melnikov, D.E., Shevtsova, V.: Experimental evidence of thermovibrational convection in low gravity. Phys. Rev. Lett. **101**, 084501 (2008)
175. Mikami, M., Oyagi, H., Kojima, N., Wakashima, Y., Kikuchi, M., Yoda, S.: Microgravity experiments on flame spread along fuel-droplet arrays at high temperatures. Combust. Flame **146**(3), 391–406 (2006)
176. Miller, R., Liggieri, L.: Interfacial Rheology. Brill Publications, Leiden (2009). ISBN 978 90 04 17586 0
177. Miyamoto, A., Shigematsu, T., Fukunaga, T., et al.: Medical baseline data collection on bone and muscle change with space flight. Bone **22**(Suppl 5), 79S–82S (1998)
178. Mizukami, Y.: A matter of size: developmental control of organ size in plants. Curr. Opin. Plant Biol. **4**, 533–539 (2001)
179. Moes, M.J.A., Gielen, J.C., Bleichrodt, R.-J., et al.: Simulation of microgravity by magnetic levitation and random positioning: effect on human A431 cell morphology. Microgravity Sci. Technol. **23**, 249–261 (2011)
180. Monti, R., Earth Space Institute: Physics of Fluids in Microgravity. Taylor and Francis, London/New York (2001)
181. Moorman, S.J., Cordova, R., Davies, S.A.: A critical period for functional vestibular development in zebrafish. Dev. Dyn. **223**, 285–291 (2002)
182. Morfill, G.E., Ivlev, A.V.: Complex plasmas: an interdisciplinary research field. Rev. Mod. Phys. **81**, 1353 (2009)
183. Morukov, V.B., Rykova, M.P., Antropova, E.N., et al.: Indicators of innate and adaptive immunity of cosmonauts after long-term space flight to international space station. Fiziol. Cheloveka **36**, 19–30 (2010)
184. Moseyko, N., Zhu, T., Chang, H.S., et al.: Transcription profiling of the early gravitropic response in Arabidopsis using high-density oligonucleotide probe microarrays. Plant Physiol. **130**, 720–728 (2002)
185. Muller, M., Aceto, J., Dalcq, J., et al.: Small fish species as powerful model systems to study vertebrate physiology in space. J. Gravit. Physiol. **15**, P253–P254 (2008)
186. Muller, M., Dalcq, J., Aceto, J., et al.: The function of the Egr1 transcription factor in cartilage formation and adaptation to microgravity in *Danio rerio*. J. Appl. Ichthyol. **26**, 239–244 (2010)
187. NCRP: Radiation protection guidance for activities in Low-Earth orbit. NCRP Report No. 132, Bethesda (2000)

188. NCRP: Information needed to make radiation protection recommendations for space missions beyond Low-Earth Orbit. NCRP Report No. 153, Bethesda (2006)
189. Nikolayev, V.S., Chatain, D., Garrabos, Y., Beysens, D.: Experimental evidence of the vapour recoil mechanism in the boiling crisis. Phys. Rev. Lett. **97**, 184503 (2006)
190. Nusgens, B.V., Chometon, G., Guignandon, A., et al.: Role of the RhoGTPases in the cellular receptivity and reactivity to mechanical signals including microgravity. J. Gravit. Physiol. **12**, 269–270 (2005)
191. Oganov, V.S., Grigoriev, A.I., Voronin, L.I., et al.: Bone mineral density in cosmonauts after flights lasting 4.5–6 months on the Mir orbital station. Aviakosm. Ekolog. Med. **26**, 20–24 (1992)
192. Olson-Francis, K., Cockell, C.S.: Experimental methods for studying microbial survival in extraterrestrial environments. J. Microbiol. Meth. **80**, 1–13 (2010)
193. Oyagi, H., Shigeno, H., Mikami, M., Kojima, N.: Flame-spread probability and local interactive effects in randomly arranged fuel-droplet arrays in microgravity. Combust. Flame **156**(4), 763–770 (2009)
194. Paa, W., Triebel, W., Eigenbrod, C., Larionov, M., Giesen, A.: The "Advanced Disk Laser" - An onboard laser diagnostics system for drop tower experiments. Microgravity Sci. Technol. **17**(3), 71–74 (2005)
195. Paille, C., Albiol, J., Curwy, R., et al.: FEMME: a precursor experiment for the evaluation of bioregenerative life support systems. Planet. Space Sci. **48**, 515–521 (2000)
196. Paloski, W.P., Oman, C.M., Bloomberg, J.J., et al.: Risk of sensory-motor performance failures affecting vehicle control during space missions: a review of the evidence. J. Gravit. Physiol. **15**, 1–29 (2008)
197. Parnell, J., et al.: Preservation of organic matter in the STONE 6 artificial meteorite experiment. Icarus **212**, 390–402 (2011)
198. Passerone, A., Liggieri, L. and Ravera, F.: Interfacial Phenomena. In: Monti, R., Earth (ed) Space Institute: Physics of Fluids in Microgravity. Taylor and Francis, London/New York (2001) pp. 46–77; Legros, G., Joulain, P., Vantelon, J.P., Fuentes, A., Bertheau, D. and Torero, J. Soot volume fraction measurements in a three-dimensional laminar diffusion flame established in microgravity. Combust. Sci. Technol. **178**(5), 813–835 (2006)
199. Perbal, G., Driss-Ecole, D.: Mechanotransduction in gravisensing cells. Trends Plant Sci. **8**, 498–504 (2003)
200. Perrot-Rechenmann, C.: Cellular responses to auxin: division versus expansion. Cold Spring Harb. Perspect. Biol. **2**, a001446 (2010)
201. Plaut, K., Maple, R.L., Wade, C.E., et al.: Effects of prenatal spaceflight on vestibular responses in neonatal rats. J. Appl. Physiol. **89**, 2318–2324 (2003)
202. Pollard, E.C.: Theoretical studies on living systems in the absence of mechanical stress. J. Theor. Biol. **8**, 113–123 (1965)
203. Pompeiano, O., Balaban, E., d'Ascanio, P., et al.: Gene expression in autonomic areas of the medulla and the central nucleus of the amygdala in rats during and after space flight. Neuroscience **124**, 53–69 (2004)
204. Pompeiano, O., d'Ascanio, P., Centini, C., et al.: Gene expression in rat vestibular and reticular structures during and after space flight. Neuroscience **114**, 135–155 (2002)
205. Rahmann, H., Anken, R.H.: Gravity related research with fishes–perspectives in regard to the upcoming International Space Station, ISS. Adv. Space Res. **30**(4), 697–710 (2002)
206. Rahmann, H., Hilbig, R., Flemming, J., et al.: Influence of long-term altered gravity on the swimming performance of developing cichlid fish: including results from the 2nd German Spacelab Mission D-2. Adv. Space Res. **17**, 121–124 (1996)
207. van Ras, N., Krooneman, J., Ogink, N., et al.: Biological air filter for air-quality control. In: Wilson, A. (ed.) Microgravity Applications Programme: Successful Teaming of Science and Industry (ESA SP-1290 ESTEC), pp. 270–280. ESA Publications Division, Noordwijk (2005). ISBN 92-9092-971-5, 2005

208. Reitz, G.: Characteristics of the radiation field in low Earth orbit and in deep space. Z. Med. Phys. **18**, 233–243 (2008)
209. Renn, J., Seibt, D., Goerlich, R., et al.: Simulated microgravity upregulates gene expression of the skeletal regulator Core binding Factor [alpha]1/Runx2 in Medaka fish larvae in vivo. Adv. Space Res. **38**, 1025–1031 (2006)
210. Renn, J., Winkler, C., Schartl, M., et al.: Zebrafish and medaka as models for bone research including implications regarding space-related issues. Protoplasma **229**, 209–214 (2006)
211. Rettberg, P., Rabbow, E., Panitz, C., Horneck, G.: Biological space experiments for the simulation of Martian conditions: UV radiation and Martian soil analogues. Adv. Space Res. **33**, 1294–1301 (2004)
212. Riley, D.A., Thompson, J.L., Prippendorf, B., et al.: Review of spaceflight and hindlimb suspension unloading induced sarcomere damage and repair. Basic Appl Myol. **5**, 139–145 (1995)
213. Rittweger, J., Felsenberg, D.: Bed-rest induced bone loss continues after re-ambulation in humans. In: Proceedings of the Human Physiology Meeting, p. C43. Physiological Society, King's College, London (2004)
214. Rivolier, J.: L'Homme dans l'espace: une approche psycho-écologique des vols habitués. PUF, Paris (1997)
215. Ronney, P.D.: Near-limit flame structures at low Lewis number. Combust. Flame **82**(1), 1–14 (1990)
216. Ronney, P.D.: Understanding combustion processes through microgravity research. Sym. (Int.) on Combust. **27**(2), 2485–2506 (1998)
217. Saint-Jalmes, A., Marze, S., Ritacco, H., Langevin, D., Bail, S., Dubail, J., Roux, G., Guingot, L., Tosini, L., Sung, P.: Diffusive liquid transport in porous and elastic materials: the case of foams in microgravity. Phys. Rev. Lett. **98**, 058303 (2007)
218. Sallam, E.A., Schmäh, M., Horn, E.R.: Hardware development for electrophysiological long-term studies in space. J. Gravit. Physiol. **12**, P271–P272 (2005)
219. Salnitski, V.P., Dudukin, A.V., Johannes, B.: Evaluation of operator's reliability in long-term isolation (the "pilot"-test). In: Baranov, V.M. (ed.) Simulation of Extended Isolation: Advances and Problems, pp. 30–50. Slovo, Moscow (2001)
220. Salter, D.M., Heißelmann, D., Chaparro, G., van der Wolk, G., Reißaus, P., de Kuyper, E., Tuijn, P., Dawson, R.W., Hutcheon, M., Drinkwater, G., Stoll, B., Gebauer, K., Molster, F.J., Linnartz, H., Borst, G., Fraser, H.J., Blum, J.: A zero-gravity instrument to study low velocity collisions of fragile particles at low temperatures. Rev. Sci. Instrum. **80**, 074501 (2009)
221. Sánchez Tarifa, C., Salvá J.J., López Juste, G.: Flame Spreading over Solid Fuels at Microgravity conditions Results Obtained in the Minitexus Rocket and Future Programmes. European Space Agency (Special Publication) ESA SP (vol. 385), pp. 281–288, ESA Publications Division, Noordwijk, Netherlands (1996)
222. Sancho, L., et al.: Lichens survive in space: results from the 2005 LICHENS experiment. Astrobiology **7**, 433–454 (2007)
223. Sandal, G.M., Bergan, T., Warnche, M.,Værnes, R., Ursin, H.: Psychological reactions during polar expeditions and isolation in hyperbaric chambers. Aviat Space Env Med **67**, 227–234 (1996)
224. Sandal, G.M., Bye, H.H., van de Vijver, F.: Personal values and crew compatibility. Results from a 105 days simulated space mission. Acta Astronautica **69**(3–4), 141–149 (2011)
225. Sauer, J., Hockey, G.R.J., Wastell, D.: Maintenance of complex performance during a 135-day spaceflight simulation. Aviat Space Env Med **70**, 236–244 (1999)
226. Schutz, B.F.: Fundamental Physics in ESA's Cosmic Visions Pan (PDF), ESA Publication SP-588 (2005)
227. Sebastian, C., Eßeling, K., Horn, E.: Altered gravitational forces affect the development of the static vestibuloocular reflex in fish (*Oreochromis mossambicus*). J. Neurobiol. **46**, 59–72 (2001)
228. Seibert, G., et al.: A world without gravity – Research in space for health and industrial processes. European Space Agency (2001)
229. Selye, H.: A syndrome produced by diverse nocuous agents. Nature **138**, 30–32 (1936)

230. Servotte, S., Zhang, Z., Lambert, Ch, et al.: Establishment of stable human fibroblast cell lines constitutively expressing active Rho-GTPases. Protoplasma **229**(2–4), 215–220 (2006)
231. Shayler, D.A.: Disasters and Accidents in Manned Spaceflight. Praxis, Chichester (2000)
232. Shimada, N., Moorman, S.J.: Changes in gravitational force cause changes in gene expression in the lens of developing zebrafish. Dev. Dyn. **235**, 2686–2694 (2006)
233. Shimada, N., Sokunbi, G., Moorman, S.J.: Changes in gravitational force affect gene expression in developing organ systems at different developmental times. BMC Dev. Biol. **5**, 10 (2005)
234. Sibonga, J.D., Evans, H.J., Sung, H.G., et al.: Recovery of spaceflight-induced bone loss: bone mineral density after long-duration missions as fitted with an exponential function. Bone **41**, 973–978 (2007)
235. Sievers, A., Buchen, B., Volkmann, D., Hejnowicz, Z.: Role of cytoskeleton in gravity perception. In: Lloyd, C.W. (ed.) The Cytoskeletal Basis of Plant Growth and Form, pp. 169–182. Academic, London (1991)
236. Slenzka, K., Appel, R., Rahmann, H.: Development and altered gravity dependent changes in glucose-6-phosphate dehydrogenase activity in the brain of the cichlid fish *Oreochromis mossambicus*. Neurochem. Int. **26**, 579–585 (1995)
237. Smith, A.M., Heer, M.H.: Calcium and bone metabolism during space flight. Nutrition **18**, 849–852 (2002)
238. Smith, S.M., Zwart, S.R., Block, G., et al.: The nutritional status of astronauts is altered after long-term space flight aboard the International Space Station. J. Nutr. **135**, 437–443 (2005)
239. Space Studies Board: Assessment of Directions in Microgravity and Physical Sciences Research at NASA. The National Academies of Press, Washington, DC (2003)
240. Stanga, J., Baldwin, K., Masson, P.: Joining forces: the interface of gravitropism and plastid protein import. Plant Signal Behav. **4**, 933–941 (2009)
241. Stein, T.P.: Nutrition in the space station era. Nutr. Res. Rev. **14**, 87–117 (2001)
242. Stowe, R.P., Sams, C.F., Pierson, D.L.: Effects of mission duration on neuroimmune responses in astronauts. Aviat. Space Environ. Med. **74**, 1281–1284 (2003)
243. Stout, S.C., Porterfield, D.M., Briarty, L.G., et al.: Evidence of root zone hypoxia in Brassica rapa L. grown in microgravity. Int. J. Plant Sci. **162**, 249–255 (2001)
244. Streb, C., Richter, P., Ntefidou, M., et al.: Sensory transduction of gravitaxis in Euglena gracilis. J. Plant Physiol. **159**, 855–862 (2002)
245. Suedfeld, P.: Groups in isolation and confinement: Environments and experiences. In: Harrison, A.A., Clearwater, Y.A., McKay, C.P. (eds.) From Antarctica to Outer Space. Springer, New York (1991)
246. Suedfeld, P.: Invulnerability, coping, salutogenesis, integration: four phases of space psychology. Aviat Space Env Med **76**, B61–B66 (2005)
247. Sundaresan, A., Pellis, N.R.: Cellular and genetic adaptation in low-gravity environments. Ann. NY Acad. Sci. **1161**, 135–146 (2009)
248. Thomas, H., Morfill, G.E., Demmel, V., Goree, J., Feuerbacher, B., Möhlmann, D.: Plasma Crystal: Coulomb Crystallization in a Dusty Plasma. Phys. Rev. Lett. **73**, 652–655 (1994)
249. Tikhomirov, A.A., Ushakova, S.A., Kovaleva, N.P., Lamaze, B., Lobo, M., Lasseur, C.: Biological life support systems for a Mars mission planetary base: Problems and prospects. Adv. Space Res. **40**, 1741–1745 (2007)
250. Tomi, L., Kealey, D., Lange, M., Stefanowska, P., Doyle, V.: Cross-cultural training requirements for long-duration space missions: Results of a survey of International Space Station astronauts and ground support personnel. Paper delivered at the human interactions in space symposium, Beijing, 21 May 2007
251. Torero, J.L., Bonneau, L., Most, J.M., Joulain, P.: The effect of gravity on a laminar diffusion flame established over a horizontal flat plate. Sym. (Int.) on Combust. **25**(1), 1701–1709 (1994)
252. Ullrich, O., Huber, K., Lang, K.: Signal transduction in cells of the immune system in microgravity. Cell Commun Signal **6**, 9 (2008)

253. van Vaerenbergh, S., Legros, J.C.: Diffusion and thermodiffusion in microgravity. In: Monti, R. (ed) Earth Space Institute: Physics of Fluids in Microgravity. pp. 178–222. Taylor and Francis, London/New York (2001).
254. Vailati, A., Giglio, M.: Giant fluctuations in a free diffusion process. Nature **390**, 262–265 (1997)
255. Vatsa, A., Smit, T.H., Klein-Nulend, J.: Mechanobiology of bone: from cell to organ. In: Monici, M., Van Loon, J., (eds.) Cell Mechanochemistry, Biological Systems and Factors Inducing Mechanical Stress, pp. 131–152. Transworld Research Network, Trivandrum (2010). ISBN 978-81-7895-458-5
256. Versari, S., Villa, A., Bradamante, S., Maier, J.A.: Alterations of the actin cytoskeleton and increased nitric oxide synthesis are common features in human primary endothelial cell response to changes in gravity. Biochim. Biophys. Acta **7**, 1645–1652 (2007)
257. Versari, S., Villa, A., Helder, M.N., et al.: Effects of gravity on proliferation and differentiation of adipose tissue-derived stem cells. J Gravitat Physiol **14**, 127–128 (2007)
258. Vico, L., Collet, P., Guignandon, A., et al.: Effects of long-term microgravity exposure on cancellous and cortical weight-bearing bones of cosmonauts. Lancet **355**, 1607–1611 (2000)
259. Vladimirov, S.V., Ostrikov, K., Samarian, A.A.: Physics and Application of Complex Plasmas. World Scientific Press, Singapore (2005). ISBN 978-1-86094-572-4
260. Vreeburg, J.P.B., Veldman, A.E.P.: Transient and sloshing motions in an unsupported container In: Monti, R. (ed) Earth Space Institute: Physics of Fluids in Microgravity. Taylor and Francis, London/New York (2001). pp. 293–321
261. Vukanti, R., Mintz, E., Leff, L.: Changes in gene expression of E. Coli under conditions of modelled reduced gravity. Microgravity Sci. Technol. **20**, 41–57 (2008)
262. van der Waarde, J.J., van Ras, N.J.P., Ogink, N. et al: Biological airfilter for air quality control of life support systems in closed environments. Proceedings of the 35th COSPAR Scientific Assembly. 18 – 25 July, Paris, France., p. 4399 (2004)
263. Wade, C.E.: Responses across the gravity continuum: hypergravity to microgravity. Adv. Space Biol. Med. **10**, 225–245 (2005)
264. Walther, D.C., Fernandez-Pello, A.C., Urban, D.L.: Space shuttle based microgravity smoldering combustion experiments. Combust. Flame **116**(3), 398–414 (1999)
265. Wang, H., Zheng, H., Sha, W., et al.: A proteomic approach to analyzing responses of Arabidopsis thaliana callus cells to clinostat rotation. J. Exp. Bot. **57**, 827–835 (2006)
266. Waters, G., Dixon, M.A., Masot, A. et al.: Static mass balance studies of the MELiSSA pilot plant. In: Integration of a Higher Plant Chamber. 34th International Conference on Environmental Systems. Colorado Springs, CO, USA, SAE paper 2004-01-2579, SAE International USA. (http://papers.sae.org/2004-01-2579, DOI: 10.4271/2004-! 01-2579)
267. Wieland, P.O. Living together in space: The design and operation of the life support systems on! the Inte vol-1 NASA Technical Manual TM-1998–206956 NASA Administration. Marshall Space Flight Center. Huntsville Al. USA (1998)
268. Wiesel, T.: Postnatal development of the visual cortex and influence of environment. Nature **299**, 583–591 (1982)
269. Williams, F.A.: Materials and fluids under low gravity. In: Proceedings of the IXth European Symposium on Gravity Dependent Phenomena in Physical Sciences. Springer, Berlin (1995)
270. Wilson, J.W.: Media ion composition controls regulatory and virulence response of Salmonella in spaceflight. PLoS ONE **3**, r3923 (2008). doi:10.1371/journal.pone.0003923
271. Wurtman, R.J., Wurtman, J.J.: Carbohydrates and depression. Sci. Am. **260**, 68–75 (1989)
272. Yates, B.J., Kerman, I.A.: Post-spaceflight orthostatic intolerance: possible relationship to microgravity-induced plasticity in the vestibular system. Brain Res. Rev. **28**, 73–82 (1998)
273. Young, S.N., Smith, S.E., Pihl, R.O., et al.: Tryptophan depletion causes a rapid lowering of mood in normal males. Psychopharmacology **87**, 173–177 (1985)
274. Zappoli, B., Bailly, D., Garrabos, Y., Le Neindre, B., Guenoun, P., Beysens, D.: Anomalous heat transport by the piston effect in supercritical fluids under zero gravity. Phys. Rev. A **41**, 2264–2267, (1990); Onuki, A., Hao, H., Ferrell, R.A.: Fast adiabatic equilibration in a single-component fluid near the liquid-vapor critical point. Phys. Rev. A **41**, 2256 (1990)

275. Zhang, Z., Lambert, Ch, Servotte, S., et al.: Effects of constitutively active GTPases on fibroblast behavior. Cell. Mol. Life Sci. **63**, 82–91 (2006)
276. Dachev T.P.: Characterization of the near Earth radiation environment by Liulin type spectrometers. Advances in Space Research. **44**, 1441–1449 (2009).
277. Rabbow, E. et al.: EXPOSE, an Astrobiological Exposure Facility on the International Space Station – from Proposal to Flight. Origins of Life and Evolution of the Biosphere, **39**(6), 581–598 (2009)
278. Horneck, G.: Impact of microgravity on radiobiological processes and efficiency of DNA repair. Mutation Research **430**, 221–228 (1999)
279. Horn, E., Gabriel, M.: Gravity-related critical periods in vestibular and tail development in Xenopus laevis. J. Exp. Zool., in press (2011)

Chapter 4
ULISSE Access to Data

Luigi Carotenuto, Christine Cornier, David Damen, Aloïs Grimbach, Victoria Lapuerta, Giovanni Minei, and Diana Pasquariello

Abstract This section describes the whole life cycle of space data, from the initial design of the experiment to the storage of data once the experiment has been performed. The need to subsequently preserve and exploit these data is also discussed; the relevant information required for data exploitation in the future is described with the main legal issues and possible data dissemination constraints. Then, the Topic Maps technology and the main features of the ULISSE knowledge base are described. The section ends with a description of the ULISSE data exploitation services, illustrated with an example of access to data through ULISSE.

4.1 Data Description

Christine Cornier[1], Victoria Lapuerta[2]

[1]Centre spatial de Toulouse, 18 avenue Edouard Belin, 31401 Toulouse Cedex 9, France
[2]Universidad Politécnica de Madrid, 28040 Madrid, Spain

4.1.1 Preservation of Space Data

The data concerned result from the experiments or observations performed by various payloads and instruments hosted on platforms, and characterise relevant parameters which are applicable to the scientific topics concerned. These data are true historical events, since the observations and measurements they represent can never be repeated: the measuring instrument, the environmental and space-time conditions and the data-processing techniques all affect any result and make it unique, and thus worthy of being preserved and accessible in the future.

The life cycle of space data is quite complex, not only due to the difficulties inherent to any scientific project, but also to the organisation and coordination

D. Beysens et al. (eds.), *Laboratory Science with Space Data*,
DOI 10.1007/978-3-642-21144-7_4,

required between the different partners involved: scientists, hardware manufacturers, USOCs, space agencies, etc. These data are then detailed: how they are produced, their main characteristics and technical composition, and the formats and scenarios used for storing media. The associated restrictions on further use in other scientific studies will have a strong impact on the availability of the data sets concerned.

The long-term preservation of data is the main purpose of this description, and its context, motivations and approach appear to be identical to other fields, aiming at a wider distribution to the scientific community and valorising these scientific results. This requires common standards to be drawn up and complied with.

On the whole, the needs and objectives of data description and classification are clarified, and the resulting data better understood.

4.1.2 Main Space Databases

The databases currently available on space experiments in the microgravity environment can be split into two categories:

- Databases which allow access to basic scientific results such as data sets:
  Currently, these are generally managed in-house by the organisations or scientific laboratories concerned, have restrictions on access, and are seldom made widely available through the Internet. None are currently well-known or referenced to provide actual scientific data sets for these experiments in the microgravity environment. The ULISSE system would provide a real improvement in this area.
- Databases which describe and detail the experiments or missions:
  These may be available and accessible on the Internet, but others are not and are kept on intranets. The descriptions follow a more or less standardised framework (metadata), and the experiment records often include references to related scientific publications or papers.

  The following databases offer such catalogues and references:

- The Erasmus Experiment Archive (EEA) [4] is a database of ESA-funded or co-funded experiments covering a wide range of scientific areas, performed during missions and campaigns on/in various space platforms and microgravity ground-based facilities. Its initial and former version was the ESA Microgravity Database (MGDB), now included within EEA.
- The GRC/NASA MICrogravity Research EXperiments Database (MICREX) [15] is a database that contains both Experiment Data Management Plans (EDMP) and related papers from experiments about Combustion, Fluids and Acceleration Measurement in the Microgravity environment.
- The JAXA International Space Environment Utilization Research Data Base (ISRDB) [13] provides results of space experiments, investigations, scientific papers, and related activities conducted by JAXA.

- The Microgravity International Distributed Experiment Archive (IDEA) is a joint international effort bringing together several distributed archive systems through a mutually agreed standard for experiment records into a common search environment. It is made up of a combination of microgravity flight experiment records contained in NASA's MICREX, ESA's EEA and JAXA's ISRDB archives. This allows users to browse through the experiments and records in these different databases, regardless of the website and starting point of the search. One objective was to develop and integrate more international partnerships.
- The Life Sciences Data Archive (LSDA) is a NASA database providing information and data from flights and flight analogue studies (ISS, Shuttle, bed rest studies, etc.) involving human, plant and animal subjects.
- The Principal Investigator Microgravity Services (PIMS) project contains the ISS Acceleration Data Archives, obtained from acceleration measurement systems (MAMS, SAMS, JAXA's MMA).
- Some Russian "catalogues" provide exhaustive lists and descriptions of microgravity experiments, such as those from:
  - OAO Rocket and Space Corporation Energia [18],
  - RKA's Scientific and Technical Advisory Council (STAC) for Programs of Scientific and Applied Research and Experiments on Manned Space Complexes [21].

Other scientific databases exist, which deal with other scientific fields and mission types, where scientists actively cooperate and share data. For example:

- a major one is the Planetary Data System (PDS), an archive of data products from NASA planetary missions, sponsored by the NASA Science Mission Directorate. This database describes the two levels mentioned previously, missions and experiments, as detailed data sets.
- the European Southern Observatory (ESO) archive, a member of the International Virtual Observatory Alliance (IVOA), is an ESO/ST-ECF science archive for astronomical observations.

The various international agencies well understood this need for preservation and valorisation of all results from scientific missions, and thus new projects appeared in various scientific fields for long-term preservation, as the European Union projects CASPAR (Cultural, Artistic, and Scientific knowledge for Preservation, Access and Retrieval) and INSPIRE.

4.1.3 Preservation and Valorisation Approach

In the overall international context, there is great demand for preservation and dissemination of data from all space missions, due to:

- the (huge) increase in the quantity of digital information resulting from missions,
- the growing complexity of scientific missions, and the difficulties and costs involved in reproducing them and their results,
- the necessary valorisation and (financial) optimisation of observations and results already obtained from these missions,
- the systematic digitisation of existing paper documents, in particular in archives and cultural fields outside the scientific domain,
- the high demand from scientific teams for interoperability and cross-reference capabilities for further and more advanced studies.

Therefore, space data, and consequently those from the USOCs, should be preserved for a very long time, to benefit all research. The principles followed for preservation should answer the question: "What is required to preserve and maintain long-term access to digital information?"

Digital preservation and dissemination systems aim at allowing access to, and interpretation of, digital data in the distant future by providing:

1. bit preservation: the ability to restore bits, and so to prevent or control deterioration and obsolescence of storage media,
2. logical preservation: to preserve future comprehensibility and usability of data, when key people, current technologies, data management products and applications, operating systems or computer hardware may no longer exist.

One important principle also influences the set of scientific data and information to be built up: scientific results in the form of data sets can be preserved and further valorised if, and only if, all the information necessary to understand the production process and detailed structure of these data, and the associated documentation and/or information, is described and linked together.

The concepts and methodologies for preservation of these digital data are formalised through the definition of standards such as OAIS (Open Archive Information System) [17] or PAIMAS (Producer-Archive Interface Methodology Abstract Standard) [19] from CCSDS [1]. These guide the following data description approach.

4.1.4 Initial Production of Data

Space experiments in the microgravity environment are conducted according to specific principles. The various types of data and results follow production cycles which depend on the technical instruments and equipment involved, the organisation of the experiment concerned, the planning of the observations during this experiment, the types of information concerned (documents, data...), etc. Some information is provided about the organisation and constraints for such experiments in ESA documents, such as the European User Guide to Low-Gravity Platforms.

The aspects detailed below directly affect the availability of data for other partners and further studies.

4.1.4.1 Organisation

The organisation and responsibilities of the partners involved in the experiment follow the experiment's main phases, i.e. the development, integration and operational (exploitation) phases. As a rule, the partners involved are:

- The experiment sponsor, generally a space agency (such as ESA, NASA, CNES, DLR, ASI, etc.), which defines the space mission based on the scientific proposals received from calls for projects and according to their Scientific Program and Programmatic Policy.
- The scientific investigator and the team (PI, Principal Investigator, and co-PI's) who proposed the experiment, who will specify and/or develop the payload and instruments, and collaborate with the sponsor during all phases of the project, from development to exploitation,
- The USOC concerned (see [5] and [8]), which is involved in the management of the experiment, and which can also manage the development of an integrated module as a framework,
- Payload development partners: space agencies, USOCs, manufacturers or companies as possible contractors,
- Operational teams from the USOCs concerned, in cooperation with the PI's scientific team in charge of operations for the experiment through distributed ground segments (MCCs for C&C of the ISS, ESA TTC stations, USOCs for scientific operations, etc.). Ultimately, the USOC is the custodian of the data resulting from the experiments concerned, and is under a contractual obligation with ESA for Long Term Archiving.

4.1.4.2 Phases of an Experiment

The data resulting from an experiment and worthy of preservation are produced through the following phases:

- development phase of the experiment, its payload, scientific and data processing tools, etc.: mainly documentation (experiment and technical specifications, test procedures, operational procedures, etc.), reference configurations, processing software, etc.,
- test phases, preliminary and complementary phases and/or within other environments for tests outside the formal sequence of the experiment: reference data sets for further scientific analysis or algorithm developments,
- pre-operational and post-operational phases following a formal schedule before and after the actual flights: reference data sets for the studies concerned; these include, for example, critical steps for experiments on human physiology,

- operational phases during sessions and runs of one or more flights in the actual "on-board" micro-gravity environment: actual in-flight results and data sets,
- scientific research and studies based on the low-level data obtained, to final status and publications.

The previously-established organisation frameworks and these clearly identified phases lead to information being distributed among the partners involved: for example, for an ISS experiment, technical specifications and documents produced by the space industry, operational products and procedures from USOCs, scientific descriptions from investigators.

4.1.4.3 Scientific Data Production Cycle

Figure 4.1 presents the main principles of the currently operational data production cycle, from raw observations to those eventually provided to scientists, where:

MCC: ISS Mission Control Centre (NASA)
COL-CC: Columbus Control Centre (DLR/ESA, Germany)
USOC: User Support and Operations Centre
UHB: User Home Base (Access from PI to TM & TC, and USOC archive)

The raw data from observations and measurements are obtained from the instruments and equipment on board the platform (ISS, sounding rocket...).

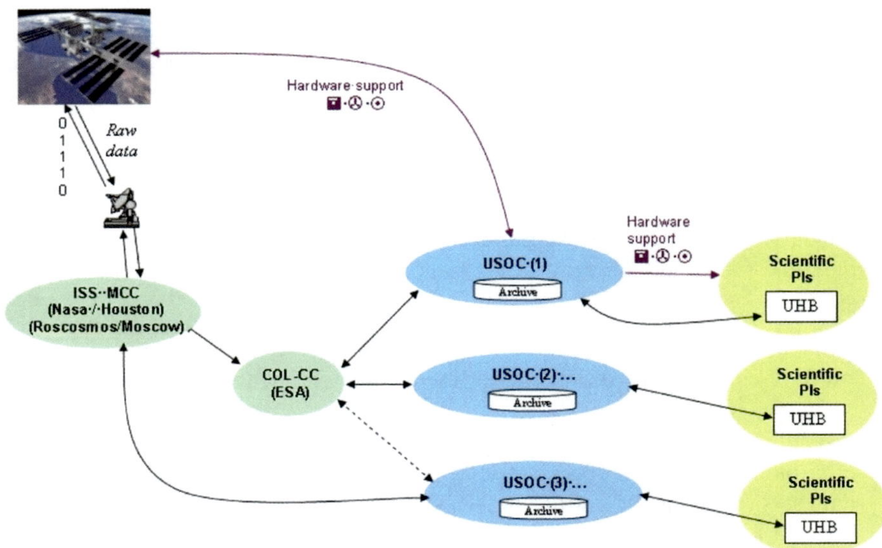

Fig. 4.1 Data production cycle for the ISS

According to the experiment definition, the data from the platform, payload and instruments concerned can be:

- transferred through the RF telemetry link to the ground segment through the TTC stations and MCC and/or COL-CC centres.
- or saved directly by the on-board payload on a local hardware medium, to be brought back to the ground with the astronauts or platform. This occurs depending on the definition of the experiment and its payload, or when the data production rate is too high for downloading.

The transferred data can be Low-Rate Data (LRD) or Medium-Rate Data (MRD), such as scientific, engineering or environment measurements (see Sect. 4.1.6) which are transferred and displayed in real time, or High-Rate Data (HRD), downloaded off-line, such as some images downloaded in "semi-real" time to check that the experiment is being performed correctly.

The flow of telemetry is processed within the Control centres (MCC, COL-CC) and sent to the USOCs, which produce the data files or databases which are then provided to PI scientists or available through UHBs.

The direct interface between the MCC and the USOCs (as shown in the diagram) concerns experiments organised as a bilateral cooperation with NASA, from the Columbus European Module.

Irrespective of the data production rate, all data sets such as data files and databases are archived and preserved within the custodian USOC, provided to PI scientists for further analysis and studies, and will be available for dissemination.

The higher-level data may or may not be provided from the PIs to the custodian USOC, depending on associated data policies and intellectual property issues (see Sect. 4.1.9).

4.1.5 Data Support and Storage

Telemetry (Low- and Medium-Rate) is archived in the COL-CC (together with other data such as Telecommands and Processed Data) for a period of 15 years, meaning that all the housekeeping and raw scientific telemetry from any payload that operated on board Columbus can be retrieved over the 15 years following performance of the experiment.

However, the SAN (Storage Area Network) in COL-CC only provides a 72-h buffer for High-Rate Telemetry; hence the HR data are decentralised at the USOCs, and long-term archiving of High-Rate data becomes the responsibility of the USOC concerned.

One of the main storage media used by the USOCs is a special hardware component called the HRDP (High-Rate Data Processor), which is a rack with servers, network equipment, the storage system and a backup system. It can store and retrieve the following data:

- High-Rate science data from the COL-CC and payloads.
- Video data from the COL-CC and payloads

- Low- and Medium-Rate telemetry from the CD-MCS.
- Processed data from and to the CD-MCS.

Nevertheless, every USOC will manage the data in a different way, depending on several factors, such as the responsibilities defined in the Experiment Scientific Requirements (ESR) document, the usual processing of the received data and the generation of final products for the Principal Investigator (PI), etc., focusing on the need to define one custom data storage policy per centre.

All of the above is applicable to the data downloaded from the ISS using the communication infrastructure, but there is also another way of transmitting the on-board data produced by the experiment directly to the ground, namely via space-craft download. Hard disks, biological samples, cells, DVDs, tapes, among other scientific assets on board the ISS, may be downloaded together with the crew, and the USOC shall be the custodian of this component.

The USOCs shall have all the scientific data classified (in terms of storage) by experiment/payload for which they have given support, for a period of 10 years: multimedia (video/audio) records, databases, images, raw telemetry, processed telemetry, file statistics, etc.

Besides the HRDP, the USOCs may use and manage other local archive systems based on storage media and technologies to preserve and share scientific data:

- Optical devices such as CDs/DVDs.
- Magnetic tapes (commonly for backup purposes)
- External Hard Disks and/or USB devices.
- NAS (Network Attached Storage) and/or File Servers.

It is also important to highlight the need for backups as part of the data storage policy, in order to be able to restore all the data in case of loss. The recommended practice is to send these backups to a secure remote location outside the USOC main building (as prevention from floods, fire or other natural disasters).

Afterwards, data are distributed to the Principal Investigator in the format and through the means previously agreed upon, applying the relevant security measures with respect to the type of data:

- Download: FTP, dropbox, WebDAV, etc.
- Shipping
- Postal system

4.1.6 Data Resulting from the Experiments and Processing Levels

The results of an experiment available as input for scientists or other potential users are:

- data grouped together as data sets, digital data satisfying the various processing levels,
- the related technical documentation,

- scientific publications, papers, presentations,
- complementary information describing the experiment,
- possible samples, depending on the scientific topic concerned, detailed in Sect. 4.1.8.

The data sets themselves can be split according to their topics and goals into:

- Scientific data, mandatory for scientists, and the main basis for further scientific studies.
- Housekeeping (or engineering) data, from the experiment container or from the facility (Fluid Science Laboratory – FSL, etc.) needed to check how the hardware worked. These data are not mandatory for scientists, but must be verified by the USOCs in real time.
- Microgravity data, considered as scientific data as they may be necessary for proper interpretation of scientific measurements, are obtained through special sensors.
- Other environmental data, sometimes useful for scientists, such as environment temperature or pressure.

For example, in Fluid Science experiments, scientific data are images or videos complemented with the measurements needed for image and video interpretation. The scientific data from the Geoflow experiment performed in the Fluid Science Laboratory (see Sect. 4.1.7) are measurements of the temperature, the rotation speed and the voltage applied to the sample, as the housekeeping data are the measurement of the current intensity in some power lines of the facility.

All of these can be produced through the different phases of the experiment.

4.1.6.1 Scientific Results and Data

Different "data processing levels", from raw and N0 to N5, are defined.

These various data levels are identified as space data (raw data, N0, N1), generally under the space agency's responsibility, correlated with the payload and instruments and the actual technical on-board environment, or as scientific data (N2, N3, N4, N5), generally under the responsibility of the PI and scientists, generated by scientific algorithms and modelling, and later used as a basis for scientific analysis and publications. The actual ownership of the data is unrelated to these levels.

These different levels of products obtained during data processing can be defined as follows:

- Raw data: Raw telemetry data from the scientific payload.
- Level 0: Data prepared ("depacketed", dated, given a chronological classification...) and quality-checked for future processing.
- Level 1: These data are derived from the level 0 data through correction by algorithms for physical or instrumental effects. They are processed in a similar way to all space data:

  - interpretation of the on-board operating process,
  - integration of transmission delays and on-board processing,
  - integration of exogenic data (orbit, attitude, weather…) necessary for scientific interpretation of the data.

- Level 2: These data are obtained from level 1 data converted into physical values, applying scientific algorithms (use of a calibration model and its associated coefficients). This covers several types of products:

  - 2A: data as physical values in digital form
  - 2B: layouts (traces)
  - 2C: summarised data ("abstracts")

- Level 3: These data are derived from level 2 data through various techniques (averaging, sampling, projection) intended to emphasise scientific acuity.
- Level 4: These data are produced by correlation or use of data from more than one experiment on the same space vehicle.
- Level 5: data are produced by correlation or use of data from more than one experiment on different space vehicles.

4.1.6.2 Documentation

As with digital data, the technical documentation is produced throughout the different phases of the experiment (see Sect. 4.1.4) in the form of different types of documents (specifications for payload and instruments, experiments, architecture, procedures, etc.).

Some of these documents appear to be essential to understanding the resulting data and being able to reprocess them. They must be linked and provided with the experiment and/or data sets concerned.

4.1.6.3 Scientific Publications, Papers and Presentations

These scientific publications in journals or at conferences are based on studies by the scientific teams, and can be considered as the initial aim and the highest level of results of such experiments, It is important to make the association between the two. These references are provided by the PIs to the USOCs.

Various means of identification are standardised and available on such systems: literary references (journal article, book), DOI link, ARK reference, link to a reference database or documentation site, link to on-line documents, etc.

Distribution of all publications is firmly restricted by copyright and intellectual property regulations, and the contents of these publications can rarely be distributed on line (see Sect. 4.1.9).

4.1.6.4 Complementary information and Metadata describing the experiment

Further study of the experiments and use of the data are possible through the system's search and access features. This requires a standardised description of the experiments, including their associated data sets and representative parameters. This significant information (such as metadata, or "data about the data") comes or is retrieved from the available documentation and knowledge, components and data files themselves. It is available through the metadata archive and/or knowledge bases. These and the metadata standard defined for the ULISSE project are detailed in Sect. 4.2.

Such a metadata scheme is generally based on a standard, ISO19115 and ISO19139 being the most commonly used. This makes it possible to provide information about:

- the metadata themselves, and the context and parameters of their production;
- the experiment: to uniquely characterise the experiment described, including identification, scientific topic, abstract, keywords, purpose for which it was performed, points of contact, scientific publications, and related documentation;
- maintenance and updates of these metadata and/or of the associated data sets;
- constraints: about the restrictions on the resource to be applied to data access and use; these may be general, legal (e.g. copyright), or security constraints (e.g. restricted data);
- acquisition: how were the data obtained, which measuring instrument mounted on which platform, along with which acquisition plan, in response to which user requirements;
- data sets: description of the data sets available for this experiment and of their contents, possibly down to data file level. The method for grouping data together into data sets is closely linked to the phases of the experiment, its runs or sessions, the source instruments, and the overall organisation.

4.1.7 Data Formats

The data resulting from the experiments can come in various formats which are closely linked with the scientific field, the source instruments and the processing level concerned (see Sect. 4.1.6). Digital data types received can be divided into:

- Images,
- Video,
- Telemetry in binary code.

At a low level, the telemetry from experiments performed in a Columbus laboratory facility (e.g. GEOFLOW experiments performed in the FSL), is received by the CD-MCS as CCSDS packets, and the data, in binary code, are stored in files according to the interface defined.

Meanwhile, the experiments performed using NASA facilities (such as SODI experiments performed in the MSG, Microgravity Science Glovebox), directly interfaced with the MCC (see Fig. 4.1), are transferred and stored in TReK format (Telescience Resource Kit).

In both cases, this telemetry is processed by the modules concerned within the USOC (CD-MCS or TReK) and databases are built from the parameters retrieved. The techniques employed to retrieve these data differ according to the module used.

Afterward, data are managed on the ground so as to be readable for scientists, and stored in files with previously agreed formats. Commonly used file formats for different data types are the following:

- Images can be stored in different file formats, such as JPEG, GIFF, EPS, TIFF,
- Videos can be stored in formats such as AVI, MPEG or MOV,
- Hypertext databases are used for parabolic flight data,
- Binary formats with respect to a standardised or well-known interface. Understanding and decoding data depends on the specific Interface Control Document (ICD) published jointly by the partners involved; for example, PSA or ECG formats for Human physiology experiments. Some generic tools are used to process data whose ICD is clearly specified.

Higher level data can be stored as:

- CSV (comma separated value), which consists of data in ASCII code separated by commas. This format is easily converted into an Excel file format,
- Excel format,
- other formats for office automation software,
- formatted text for summaries.

Other binary and proprietary formats can also be produced. These are specific to the instrument concerned, are not standardised, and the interface is unknown. In such a context, either a decoding interface, or a tool, or the source instrument itself is provided, and is necessary to process and retrieve the scientific data. In this case, only higher level data are to be disseminated.

Finally, PDF/A is the format now mainly used for documentation and publications to ensure long-term preservation.

4.1.8 Samples

Some scientific studies are based on experiments involving samples as detailed in Sect. 4.2. The life cycles of these samples can vary. Also, the associated platform determines whether or not they can be brought back on the ground for further study phases. Certain experiment processes can be performed on the ground on a control group for reference measurements. But some do not "survive" the on-board experiment, and are not significant, even if returned to the ground.

In the following scientific fields samples usually are relevant:

- Human Physiology: Biological samples, such as blood, saliva, urine, serum, tissues, etc. These are obtained through set procedures and preserved afterwards in very specific conditions.
- Animal Physiology: The relevant "samples" in this case are animals, such as rats, bees, beetles, fish, cats, monkeys, etc.
- Plant Physiology: samples are plants, seeds and sprouts of involve widely-used model plants, cultivated in space.
- Biology, Astrobiology and Biotechnology: Samples as cells, protoplasts, etc.
- Material Science: The samples usually consist of metallic alloys, composite materials, pure and compound crystals; the samples are generally resistant mechanically, with the exception of organic crystals, particle aggregations, etc.
- Technology: materials exposed to space, welding or soldering samples.

Even if samples, as results of the experiments, are interesting for other scientists, their availability is highly restricted due to the policies and European or national regulations they come under (see Sect. 4.1.9). This mostly applies to Human physiology.

The constraints on their possible transportation and then preservation, from the actual experiment environment in microgravity to the preservation facilities on the ground, will also have a strong impact on the actual transfer capabilities and conditions of some samples intended for further studies.

These various constraints must be described as part of the metadata relating to the experiment, and of the related documentation.

Currently, it is obvious that there is no real "Biobank" (for biological topics) or larger "samples bank" to properly archive these samples resulting from space experiments. Such a need is still to be considered.

4.1.9 Legal Aspects and Ethical Issues

These issues directly influence the dissemination services (for example, as implemented in ULISSE), and must lead to the classification and processing of information according to status and sensitivity (public, non-sensitive, restricted, confidential, etc.).

4.1.9.1 Responsibilities and Global Impacts

The allocation of responsibilities for activities associated with a given experiment has an impact on how the ownership of the data is spread among the partners. The actual owners of the scientific data can be the space agencies (ESA, NASA, or national agencies, etc.), mainly for raw and level 0 data (sometimes level 1), or the scientific investigators (for higher levels, 1–5). The USOCs remain the custodians of the data for the experiments they are responsible for through relationships with ESA. Neither the USOC, nor the contractor, nor its sub-contractors, nor their representatives or agents, can disseminate or disclose to third parties any data, results or information without prior authorisation from the Agency's representatives. The legal rules to be complied with and the policy-enforced restrictions affect data availability and accessibility, for example through a system like ULISSE, and therefore the on-line or off-line availability of the data sets.

These restrictions mainly concern the distribution of scientific data sets, but metadata used for data discovery must comply with them for data to be freely available to any authorised system user.

These constraints are expressed through the ESA data policy when the experiment is funded by ESA, otherwise through national policies and regulations.

4.1.9.2 ESA Data Policy

With regard to experiments mainly referenced within ULISSE, and funded or partially funded by ESA, the distribution of data owned by ESA must comply with the ESA data policy formalised in [6] and summarised in [7] i.e.:

- for Raw and Calibrated Data, resulting from the implementation of the Experiments: ESA grants the PI an exclusive right of prior access to these for 1 year, after which States, Persons or Bodies may request access to the data from the Agency by providing specifications and justifications. ESA shall grant such access after studying certain conditions (do the data come under ESA jurisdiction, what is the PI's point of view, etc.). Unless commercial interests are involved, such conditions should normally provide for unrestricted access.
- for Analysed Data: these can be produced and thus owned either by the Investigator or by the Agency (ESA); this defines the associated intellectual property and distribution procedures.

When authorised, distribution can be free of charge, or include fees for copying and other reasonable administration costs, or come under other conditions for other uses.

It is also obvious that changes are currently occurring on some of the ESA data policies, mainly for Earth Observation products concerned by the GMES project and its wider open services which could be enlarged to other fields.

4.1.9.3 Human Physiology and Ethical Issues

As with other scientific fields, the results and data sets from these experiments can be owned by space agencies or by the PI's. However, if astronauts or volunteers are involved, the distribution of these data is constrained by ethical laws or policies governed by the national regulations applicable for the USOC involved and by ESA regulations.

For example, in compliance with French law, this authorisation is determined by the scope of further studies and by the similarity of the new scope to the original one, by the submission to Ethics committee and French Health Products Safety Agency (AFSSAPS), and by formal approval from the volunteers and authorities.

4.1.9.4 Publications, Copyrights and Documentation

Although any reference to the publications relating to an experiment can and should be freely provided, distribution of the actual document must comply with intellectual property regulations and with the copyright restrictions associated with the journal or symposium. This sensitive restriction often prohibits the release of the actual publication from such a distribution system.

The technical documents provided on the experiments include those under the Agency's responsibility and funded by the Agency. This can also apply to industrial documents, provided they are free from intellectual property restrictions.

4.2 Identification of Information: The ULISSE Knowledge Base

David Damen[1], Alois Grimbach[2]

[1]Space Applications Services, Leuvensesteenweg, 325, B-1932 Zaventem, Belgium
[2]Microgravity User Support Center MUSC, Linder Höhe, D-51147 Cologne, Germany

4.2.1 Purpose and Role of Semantic Technologies

4.2.1.1 The ULISSE Metadata Standard

The aim of the ULISSE project is to provide (and enable the creation of) descriptions of experiments in space science and related fields to interested parties. The purpose of these descriptions is two-fold:

1. To describe the data captured through the execution of the experiment.
2. To capture the knowledge needed to understand the experiment.

These 'descriptions explaining data' are called metadata. The main problem with any such metadata description is its heterogeneity. Any individual providing metadata will do so from structured and unstructured data representations used in his particular knowledge domain, from his own experience and expertise, and from the perspective according to which he addresses his knowledge domain. Left unchecked, a proliferation of personalized descriptions will abound which are impossible to correlate and integrate. This risk is especially prevalent in the case of ULISSE, which brings together data from such a wide variety of scientific disciplines.

To mitigate this risk a metadata standard for ULISSE has been created. A metadata standard defines a structure that should be followed when describing an experiment. This ensures that every experiment description provides the same level of information in the same format; thereby allowing better comparison and understanding of space experiments and their data.

The metadata standard developed in ULISSE describes individual experiments and focuses on data preservation aspects. It is based on ISO 19115 and includes extensions dedicated to microgravity experimentation [14]. Accordingly, it contains sections on the experiment itself, how the data and metadata are maintained, constraints on their use, and data acquisition and contents.

4.2.1.2 Semantic Technologies

While a metadata standard provides considerable improvements in the representation of knowledge descriptions, some issues still remain. It is accompanied by a lengthy textual description on its use and, despite a rigid structure, leaves some elements open to interpretation or provides some users with a less than optimal composition, e.g. what constitutes an experiment run differs greatly depending on the experiment setup.

In order to remove those ambiguities, a *semantic* representation of the metadata standard has also been pursued. The meaning of semantic data or a semantic description was accordingly defined by David Siegel as follows "*Data that is semantic means exactly the same thing to any system or person who uses it.*" [22] In ULISSE, Topic Maps were chosen as the technology to provide the semantic representation of the metadata standard.

4.2.2 Topic Maps as a Semantic Data Model

Topic Maps are a standardized technology (ISO/IEC 13250) for representing knowledge which allows the **meaning** of the stored data to be recorded.

4.2.2.1 Topic Maps Data Model

The Topic Maps Data Model (TMDM)[1] defines a formal data model for topic maps. It is used to define the XML Topic Maps (XTM) syntax, and is the foundation for

[1]Topic Maps standard, ISO/IEC 13250-2.

the Topic Maps Query Language (TMQL) and the schema language (Topic Maps Constraint Language (TMCL)).

Topic

In Topic Maps, every subject in the knowledge domain is represented by a topic. For example, in Fig. 4.2, ***Space Applications Services*** is a topic representing the company *Space Applications Services*. Topics can represent anything, even other topics. It is the responsibility of the author of the topic map to decide what subjects will be represented as topics in the topic map. Obviously, for a more detailed topic map, more topics are defined.

Topic Type

Each topic may have a type. ***Space Applications Services*** is a company and therefore its type is ***Company***. In this case, ***Space Applications Services*** is said to be an instance of the type ***Company***. As can be seen in Fig. 4.2, the type of a topic is itself a topic.

Associations, Players and Roles

Using Topic Maps technology different topics can be associated with each other. For example, the topic ***ScienceCast*** is associated with the topic ***Space Applications Services*** and their association is of the type ***made by***. Both the topics ***ScienceCast*** and ***Space Applications Services*** are called **players** in the association.

In order to understand which topic is ***made by*** which, **roles** are defined: the topic ***Space Applications Services*** plays the role of ***Company*** in the association of type ***made by***, while the topic ***ScienceCast*** plays the role of ***product*** in the same association (Fig. 4.3). Note that the association type (***made by***) and the roles (***Company*** and ***product***) are themselves topics.

Associating topics with each other provides a structure that corresponds to the way we conceptually grasp knowledge. Such a structure provides a way to navigate between the topics in the topic map. It also lets users ask questions about the knowledge within the topic map, for example: who made ScienceCast. This gives **meaning** to the data that is kept in the topic map.

It should be noted that although topic types are special elements in Topic Maps, they are actually just players in an ***is-a*** association. For example, we could say that the topic ***Space Applications Services*** is associated with the topic ***Company*** by the association of type ***is-a,*** where the topic ***Space Applications Services*** plays the role of ***instance*** and the topic ***Company*** plays the role of ***type***.

One association type related to this is the ***superclass-subclass***. From Fig. 4.4, it can be understood that ***ScienceCast*** is an instance of type ***Web Application***, while

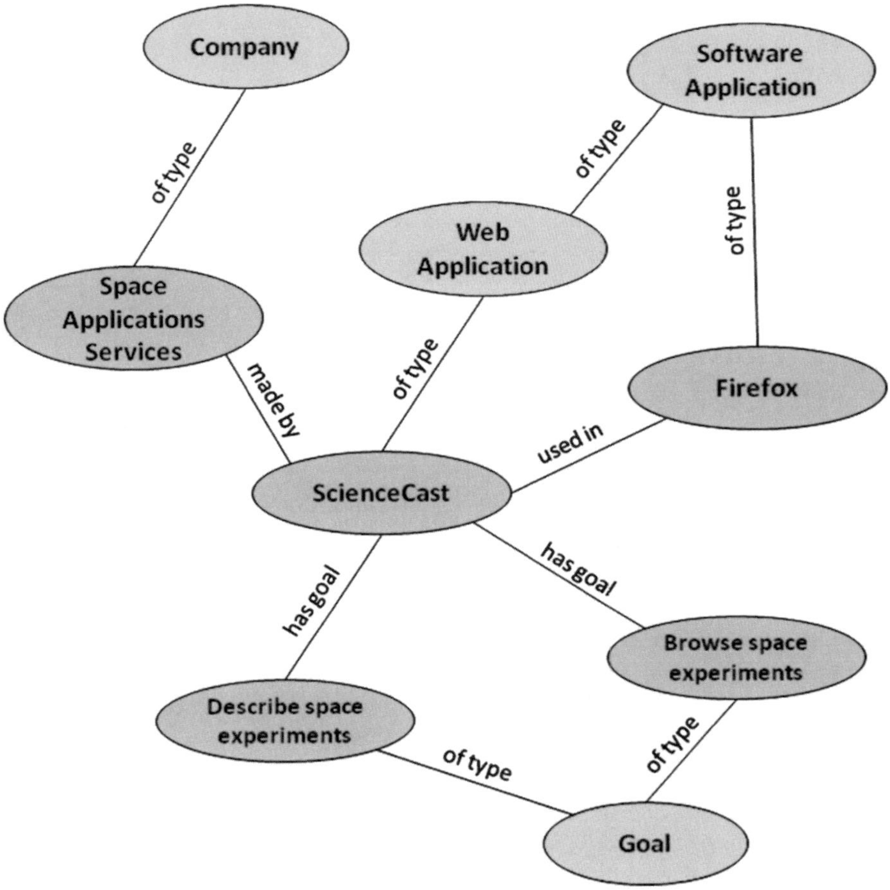

Fig. 4.2 Example of a topic map

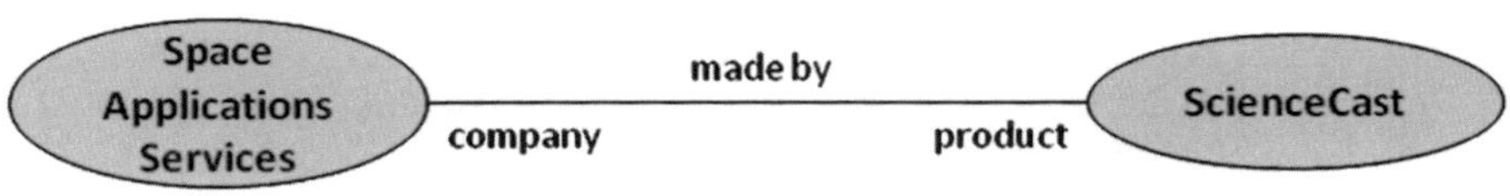

Fig. 4.3 Example of roles

Web Application is an instance of type ***Software Application***. However, the correct way to represent the relationship between those topics is as follows: ***Web Application*** is a subclass of ***Software Application*** and not an instance of it.

The ***superclass-subclass*** association type and its roles ***superclass*** and ***subclass*** are defined in the Topic Maps standard.

Finally, until now only associations between two topics were described. However, associations can be defined between more than two topics. For example, the

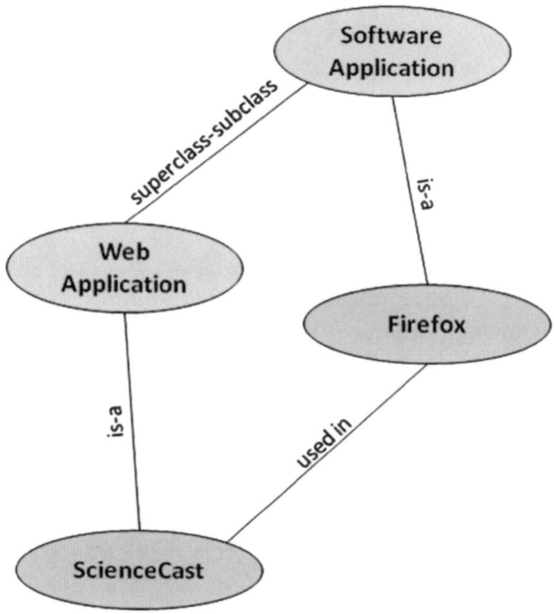

Fig. 4.4 Example of the superclass-subclass association

association ***provide-product-provided*** can be used to describe the relationship between the provider of a certain product, the product that is provided and the purchaser of that product.

Names

Each topic may have one or more names. Those names will represent the topic.

Occurrences

Each topic can hold extra information called **occurrences**. These are items of information about a topic (similar to how words in a book are referenced by the book's index). Occurrences can be simple textual information or references to any other media (such as web pages, audio, video, etc.). Therefore, the topic ***Space Applications Services*** might have as occurrences an address, a telephone number or a reference to its web page URL.

Each occurrence might have a type that defines its meaning. For example, an occurrence representing the address of a company might have the type ***address***.

Usually the topic map will not contain large occurrences, and in general, only simple, unformatted text can be stored in it. This is why two sorts of occurrences can be defined: resourceData and resourceRef. **resourceData** is textual data that is usually limited in size. **resourceRef** is a textual reference (e.g. a URL) to data that is available elsewhere. For example, the address of a company can be stored as a resourceData, however, the company overview will be referenced by providing a resourceRef that contains the URL to the relevant web page.

Scopes

Topic names, occurrences and associations can all be scoped. The scope can then be used in order to filter those elements in or out when using the topic map. One example of scoping is to scope by language. For example, a topic representing the subject *lung* may have the name "lung" scoped by the topic ***en*** (which represents the subject English) and another name "long" scoped by the topic ***nl*** (which represents the subject Dutch), thereby making the topic map multilingual.

Another use of scopes can be to describe a context in which topics might be used or understood. For example, all associations describing a logical relation between devices could be scoped by the topic ***logical*** while associations describing a physical relation between devices could be scoped by the topic ***physical***.

Ontology

A topic map is built from topics that represent specific subjects in the domain of knowledge that the topic map describes. In addition, it contains topics that represent the different types, classes, association types, occurrence types and scopes. This second group of topics can be seen as the structure, or skeleton, of the topic map. This structure is called the **Topic Maps Ontology**.

Ontologies are useful when authoring topic maps. If the ontology is defined carefully at the outset, it helps to identify the boundaries of the knowledge domain that the topic map represents. In addition, when the ontology for a certain knowledge domain is created, different topic maps can be authored using the same ontology.

Merging

One of the interesting features of Topic Maps is that they can be merged. Topic maps are merged by adding all the topics and associations of one topic map to another topic map. However, when two topics are found to represent the very same subject, those topics are merged themselves resulting in one topic which contains all the names and occurrences of both merged topics. Finally, duplications are eliminated, so a topic will not contain two identical base names or occurrences, and every two associations that relate to the very same topics are reduced to one association between those topics.

XTM, CXTM, CTM, JTM

There are several data formats for representing topic maps.

XTM[2] is an XML representation of Topic Maps, and is intended for machine exchange of information stored in topic maps.

CTM[3] is a compact textual notation for representing Topic Maps. It was originally developed to facilitate the need in TMQL to support INSERT operations, which brought forward the need for a more shorthand notation. XTM and CTM are the official serialization formats for the TMDM.

Unlike XTM and CTM, **JTM**[4] is not an industrial standard, but a community one. JTM is intended for web applications which process topic maps on the web browser, and uses JSON to represent Topic Maps.

4.2.2.2 Topic Maps Query Language

The TMQL is a standardization effort with the aim of creating a standard for a Topic Map query language. A TMQL query is processed and evaluated in the context of one or more topic maps by a TMQL processor. TMQL itself is not an implementation of the language, but the standard description. TMQL supports information retrieval and filtering, but may also be used for updating and removing data from a topic map. TMQL was inspired by research conducted in the area of information retrieval in conceptual graphs and graphical query languages developed for hypertext structures.[5]

As one of the principles when designing TMQL was to leverage existing popular querying solutions, TMQL defines a syntax closely resembling that of SQL, making it easier for developers already familiar with SQL to learn and use the language.

4.2.2.3 Topic Maps Constraint Language

TMCL[6] is a language for defining schemas and constraints on instances of the TMDM. The constraint language provides a constraint model, related validation semantics, and CTM syntax templates for authoring constraints. With TMCL it is possible to create schemas for topic maps that constrain what structures are allowed in the data set, such as:

---

[2]XTM syntax definition. http://www.isotopicmaps.org/sam/sam-xtm/.

[3]CTM ISO Standard Description. http://www.isotopicmaps.org/ctm/.

[4]JTM syntax definition. http://www.cerny-online.com/jtm/1.0/.

[5]TMQL ISO Standard description. *Robert Barta, Lars Marius Garshol*. http://www.isotopicmaps.org/tmql/.

[6]TMCL ISO Standard Description. http://www.isotopicmaps.org/tmcl/.

- "an animal must have a weight",
- "if an animal feeds off other animals, it must play the carnivore role in the ***feeds-off*** association, while the other animal must play the role of prey", or
- "an animal of type fish cannot play a role in the ***has-lungs*** association".

Through specifying these constraints TMCL provides a formal ontology definition and facilitates the validation of topic maps, where a topic map is considered valid if it conforms to all the constraints of a given schema, as well as to additional constraints that apply to all topic maps. TMCL does not have its own syntax, but is defined as part of the Topic Maps vocabulary.

4.2.3 The ULISSE Knowledge Base

The ULISSE Knowledge Base implements the TMDM and thus offers an extensible solution for adding a representation layer on top of a pool of heterogeneous datasets. It is capable of hosting and displaying interconnected knowledge structures in a scalable way, hence advancing cross-fertilization across scientific disciplines in the domain of microgravity experimentation. ULISSE hosts domain knowledge and data sets from the disciplines Material Science, Fluid Science, Space Weather and Solar Physics together with Biology, Plant Biology, Human Physiology and Respiratory Physiology.

In general, creating an Ontology is an iterative process, involving both scientific experts and knowledge engineers [10]. Each iteration aims at enhancing coverage as well as granularity of the ontology. The ontology coverage can be assessed by asking competency questions from several knowledge domains which in turn can expand the scope of the ontology.

Despite the fact that the ULISSE topic maps cover experiments from different scientific disciplines, a certain overlap became apparent during their iterative development [3]. This overlap is mostly related to how microgravity experiments are structured and set up. The intersection of all the concepts was defined as a first General Space Experiments Ontology with common identification labels. In that sense, it provided the first example of knowledge interconnection by using similar concepts in different domains.

Strategies for populating topic maps comprise direct authoring with specialized tools, the reuse of existing metadata or a combination of both. In many cases, existing metadata are available in databases or as XML files and can be converted to topic maps. Usually, it is necessary to manually update the generated topic maps to make them more usable.

In the case of ULISSE, most of the metadata had to be generated or set up manually. During this process, all the data providers supplied plain XML descriptions of their experiments which conformed to the ULISSE Metadata standard. Consequently, a ULISSE Metadata part forms part of the General Space Experiments Ontology. It contains fields about each experiment such as the abstract and

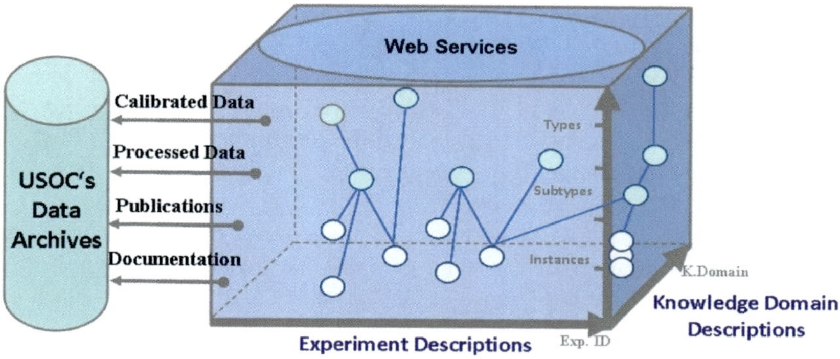

Fig. 4.5 The ULISSE knowledge base

Principal Investigator and, when applicable, about the associated scientific program. Also metadata about data maintenance, dissemination constraints and detailed information about data acquisition are included in the standard. The description of particular datasets is organized in a nested structure, entailing information about measured or imposed parameters, applied data formats or online links to the data, where available.

Figure 4.5 gives a schematic view of the ULISSE Knowledge Base. Experiment descriptions in XML format, according to the ULISSE metadata standard, serve as input to the Knowledge Base. They are located on the *Exp.ID* axis and are mapped to a populated topic map in the interior of the cube, thus forming its semantic description as part of the ULISSE Knowledge Base. The ULISSE specific conversion from plain XML files to topic maps forms a crucial part of the knowledge base architecture and demonstrates how an ISO standard can be transformed into a Topic Maps representation. On the other hand, it reveals intrinsic restrictions in the mapping from an XML Schema (XSD) to topic maps.

Other mappings are conceivable and may be incorporated into the ULISSE Knowledge Base, possibly by giving potential data providers guidelines for generating Topic Map representations. This principle may help in finding the balance between proprietary and open metadata standards, and then being able to integrate them into a joint Knowledge Base.

4.2.3.1 Knowledge Description and Development

Apart from the introduced metadata, any microgravity experiment has a scientific part whose structured description is in general not part of the mentioned metadata fields. In a non-semantic metadata standard, it can merely be described by keywords or within a fluent text description, hence remaining implicit and dependent on the context. Here, topic maps make it possible to describe and superimpose domain knowledge as an additional knowledge dimension as schematically indicated by the *K.Domain* axis in Fig. 4.5.

For example, a user wants to store the fact that a specific experiment aims at investigating the thermo-physical properties viscosity and surface tension. They may have been investigated under microgravity conditions in which the liquid sample has been stimulated to oscillate. By studying the characteristics of the oscillation, viscosity and surface tension can be determined. The user then wants to store the fact that another experiment exploits oscillating characteristics but investigates the electrical conductivity.

By means of Topic Maps, the user can assign the topic type *microgravity experiment* to both experiments. He can then generate the topics *viscosity* and *surface tension* whose topic type is *thermo-physical property* and associate them with an association of type *investigation*. He can then create the topic *oscillating drop method* (of the topic type *method*), associate it with the experiment through an association of type *application* and describe the method (through an Occurrence of type *description)*, possibly with references to publications (as resourceRefs). The second experiment can be associated with the previously generated *oscillating drop method* and with an *investigation* of the topic *electrical conductivity* whose topic type is *electrical property*.

Similarly, it is possible to refine knowledge which is already contained in the topic map. For example, a topic type *Sample* may contain an occurrence of type *Material,* denoting the composition of the sample. By transforming the Occurrence of type *Material* to a topic type and associating it with a *Sample* through an association of type *composition*, assertions can be made about the material itself. This process is known as reification [11]. Reifying new pieces of information allows for detailing and extending knowledge.

Having reified the sample materials and possibly having merged material properties (e.g. stemming from an external Material Science Ontology), the topic map is able to answer a question such as "*Which experiments used Aluminium-based materials?*" [12]

On the one hand, this opens the possibility of actively working on the Knowledge Base by populating domain-specific topic maps; on the other hand it requires support for authoring topic maps, demanding sophisticated usability. Within ULISSE, a Topic Maps authoring tool has been developed to provide a user interface that hides the complexities of the Topic Maps technology and provides smart display options. It can be used to enhance and refine the Knowledge Base, respecting its underlying Ontologies. What remains locked and unchanged in the Knowledge Base is the core part which has its counterpart in the plain description of the ULISSE metadata standard.

4.2.3.2 Data Access and Information Extraction

The Topic Maps representation serves as the basis for a set of web services which are displayed by the ULISSE API, in Fig. 4.5 on the top side of the cube. Web services provide mechanisms to let machines interoperate over a network.

The ULISSE web services are based on a resource-oriented approach which is known as Representational State Transfer (REST) [9] and proposes the use of resources and their possible representations. In a RESTful application, resources are hyperlinked together and their representations are called to interact with them through standard web protocols.

In fact, an experiment description according to the ULISSE metadata standard and its assigned topic map are both different representations of the same resource, i.e. the experiment in question. Its representation can be called either as a topic map or as ULISSE metadata conforming to the XML description, depending on the request.

The principle of displaying experiments and their representations through URIs can be extended to the display of any topic contained in the ULISSE Knowledge Base. By applying the principles of REST, this leads to a resource-oriented API to the ULISSE Knowledge Base [2]. Once authorized, web developers can make use of it and build up their own applications.

As an example application, basic searching and browsing services are demonstrated on the ULISSE Web Site,[7] making use of the REST API to the ULISSE Knowledge Base.

Concerning advanced interactions with topic maps, there are efforts ongoing to develop natural language tools for processing queries over topic maps. The idea is to use a natural language phrase in order to automatically guide a search over the topic map. As a result, a topic map fragment is displayed that most likely matches the answer. Research activities from outside ULISSE [16, 20] could be used and partly integrated into the ULISSE GUI. In most cases, the resulting topic map fragment contains collateral information and helps in detecting additional but applicable information.

Pursuing the example of the previous section, a user may ask which experiments aim to investigate *thermo-physical properties*. He may retrieve a map of associated experiments, together with the associated *samples*, whose concrete properties were investigated (i.e. *viscosity* and *surface tension*) and possibly learn about the applied *oscillating drop method* (which usefully contains an occurrence field for its description).

Similarly, the user may retrieve data sets which are associated with experiments (or possibly also other topics). The data sets are physically stored in distributed data archives at the USOCs, as shown on the left-hand side in Fig. 4.5. They can comprise raw, calibrated and processed data as well as publications or documentation. Through resourceRefs, the Knowledge Base contains information on how to access the information embedded in its contextual topic maps description.

From the outside world, data sets are visible to authorized users as resources through the API. In such a way, web services can be built which either provide direct access to the data sets or process them on-the-fly and display them as new

[7]http://www.ulisse-space.eu.

resources or in suitable representations. Due to the application of standard web transportation protocols, proven security mechanisms can be utilized to respect any kind of data policy for accessing those resources.

In an interdisciplinary view, the above used topics may contain links to various databases. Ideally, that enables the detection of coherences and the use of data across scientific disciplines, always respecting the applicable data policies. This happens on a platform which allows for inclusion of new data sets, description and reuse of old ones and generation of new information fragments and associations among them.

4.3 ULISSE: Data Access Services

Giovanni Minei[1], Diana Pasquariello[2]

[1,2]Telespazio s.p.a, Via E. Gianturco 31, 80146 Naples, Italy

4.3.1 Services for Obtaining Data

Improving the preservation, valorisation and exploitation of experimental data for further scientific uses is essential for the numerous European experimental facilities hosted on the International Space Station (ISS).

ULISSE [23] represents an innovative infrastructure, which until now has been lacking in the domain of European scientific experiments on board the ISS. ULISSE is based on the existing experimental data archives distributed in the network of USOCs, and the related metadata and knowledge about the experiments. This infrastructure aims at:

- making data accessible through services based on a semantic representation of related knowledge;
- preserving and disseminating information;
- developing data using specific tools integrated in the infrastructure.

The architecture of the infrastructure has been defined on the basis of:

- identification and analysis of the existing resources and data, defining associated services and available interfaces;
- identification of innovative classes of services and additional resources useful for implementing a cooperative approach;
- identification of the service flows for exploitation of the domain knowledge represented in the Knowledge Base.

4.3.1.1 How Is the Service Provided by ULISSE Defined?

The definition of the service provided by ULISSE is closely dependent on an operational situation.

The main elements which give meaning to ULISSE services are:

- Users,
- Resources.

The set of potential users is grouped into categories; each category is characterized by a specific user role within the domain. The identified roles are mapped to the following user categories:

- Scientists
- Space Agencies
- Teachers
- Students
- Space Industries
- High-Tech Industries
- Journalists/writers
- Physicians
- Generic users

A user belonging to a category interacts with ULISSE according to the goals related to his role (for example: a Scientist may have the goal of defining a new experiment, while a Teacher may have the goal of preparing a lesson).

In this sense, a role describes the intentional context which guides the user in his interaction with ULISSE (Fig. 4.6). A role:

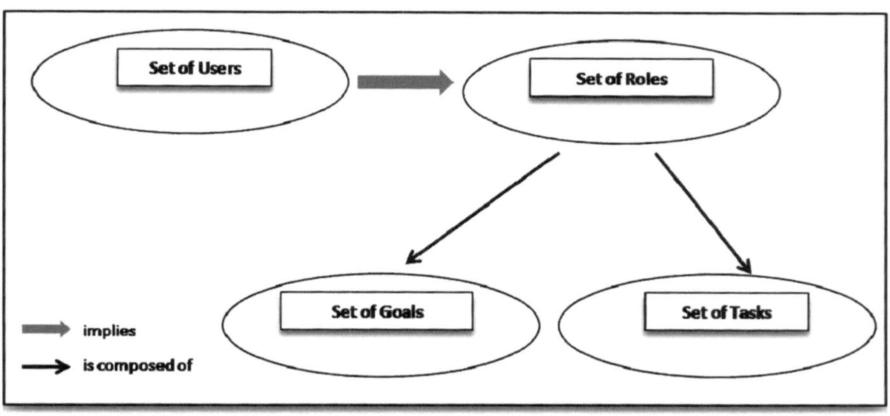

Fig. 4.6 Representation of the roles

- detects a motivation to use ULISSE;
- identifies a set of goals which come from this motivation;
- identifies a set of tasks (actions) to be performed to attain each goal.

Each example defines a particular scenario (service case). Therefore:

Any user that interacts with ULISSE has a specific role and specific goals, to be achieved in a particular scenario (service case).

Each scenario, that defines the method of interaction between the user and the ULISSE Platform according to his role and his goals, involves the resources (Fig. 4.7). In this context, a resource is made up of:

- a functionality, defined as a "basic need that motivates the user to use the resource"
- one or more sources – including the "physical link" and the "data" (e.g. databases, SW tools, file systems, web sites and portals, etc.) to be accessed.

For example, a standard data base (DB1) could be the source of different resources:

- R1 [Functionality: accessing in order to read; Source: DB1]
- R2 [Functionality: storing new data; Source: DB1]
- R3 [Functionality: deleting obsolete data; Source: DB1]

Moreover, considering another data base DB2:

- R4 [Functionality: migrating data; Source: (DB1, DB2)]

Therefore a service (Serv) can be defined by a particular ULISSE utilization:

- to obtain a set of Resources (R)
- to activate the Functionalities associated with the selected resources
- to access the Sources involved in the resources

Basic Functionalities are the key elements of the ULISSE Services.

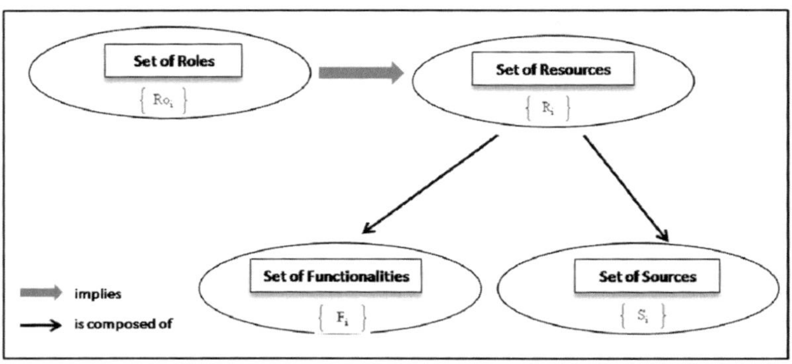

Fig. 4.7 Representation of the resources

4.3.2 *ULISSE Services*

4.3.2.1 Classification of ULISSE Services

The ULISSE domain distinguishes three main service typologies:

- Ontological Services
- Tools Services
- Security Services

Ontologies have been defined to represent the data/information related to the ULISSE informational resources. Instances of Ontological Services use the ontology defined to represent the information related to the source domain.

Tools Services are related to the available ULISSE tools including the E-collaboration portal, Data accessibility and distribution tools, Data organization & integration tools, Planning & Validation tools, Augmented Reality tools.

Security Services ensure the basic security requirements for ULISSE. These services protect information from unauthorized access. The Security Services were defined on the basis of security properties. Data access and information management play a crucial role in ULISSE: they have to be compliant not only with the ESA Data Policy but also with appropriate information security policies. Security properties identified for ULISSE data access and the derived services are briefly described below.

- Confidentiality
  Several restrictions may be imposed by the data owner on the data and information access. The International Standard ISO/IEC 17799 covers data security under the topic of information security; accordingly, an appropriate Access Control policy was defined to control access to protected information.
- Integrity
  Different actors are involved within ULISSE and their responsibilities have been clearly defined and the related data access rules have been identified in order to guarantee the Integrity property. Metadata are created by the USOC/data provider responsible for the dataset; the data provider defines an Editor responsible for creating and saving metadata; a Reviewer (also appointed by the data provider, but different from the Editor) verifies that metadata records are compliant with the ULISSE metadata standard. This process ensures users the confidence level of the information that can be gathered from ULISSE.
- Availability
  ULISSE offers connectivity between all the nodes of the network.
- Security classification for information

In line with ESA's Resolution on Information, Data and Intellectual Property, the Data Policy makes a distinction between:

- Raw and Calibrated Data, resulting from the execution of the Experiments;

- Analyzed Data, produced by either the Investigator or the Agency.

The ULISSE data policy appropriately regulates ownership, use by the Agency, intellectual property, prior access and access to, and storage of, the Raw and Calibrated Data and the Analyzed Data considering also any case of restrictions such as confidentiality and ownership rights (e.g. for medical data). ULISSE services properly address this issue case by case, considering the different legal aspects related to the specific context.

4.3.2.2 Basic Functionalities: Definition and Related Examples of Scenarios

Basic functionalities were defined starting from a set of questions identified by data providers as being representative of the needs of the user categories.

These questions have been structured into several categories:

- Questions related to Science and Terminology
- Questions related to the Experiment (setup, execution, output and data)
- Questions related to Instruments and Experimental Facilities
- Cross-Experiment Questions
- Questions related to ULISSE
- Questions related to the International Space Station (ISS)
- Miscellaneous

Every question was associated with one or more basic functionalities. Each functionality gives an explanation of "how" information regarding a question could be used by a specific user. Each Basic Functionality uses the ontology to retrieve relationships among the required information to build the answer.

The following Basic Functionalities were identified.

Knowing

This functionality fulfils the most important scope of the ULISSE platform to users: retrieving the required information. This functionality provides standard access to the sources to retrieve the existing stored information.

The KNOWING functionality is characterized by:

- access to the Metadata Server
- access to the available sources
- display of information

Every other basic functionality uses "Knowing" to retrieve information.

Practical Explanation

Starting from a specific user request for information that is not explicitly present in the associated sources, this functionality collects the available data/information, retrieved by the Knowing functionality, and composes an explanation of the subject based on specific composition rules following the ULISSE Ontology.

PRACTICAL EXPLANATION involves the following actions:

- choosing a presentation layout
- accessing the Metadata Server (using Knowing)
- accessing the available sources (using Knowing)
- composing the PRACTICAL EXPLANATION output

Scenario: a teacher asks for a Practical Explanation of Marangoni Convection, composed of textual descriptions, photos and videos. ULISSE retrieves textual descriptions, videos and photos on Marangoni Convection. Then ULISSE composes the retrieved information in the selected layout.

Matching

This functionality creates a set of matches among heterogeneous information which could be retrieved using the Knowing functionality. These matches are defined on the basis of specific correspondence rules. MATCHING involves the following actions:

- selecting keywords related to the topics to be matched
- accessing the Metadata Server (using Knowing)
- accessing the available sources (using Knowing)
- defining the matching using Correspondence Rules
- displaying the matched Information related to the keywords

Scenario: a Space Agency wants to know the equipment component activated by the commands of a given procedure. ULISSE, on the basis of the procedure steps and the H\W used in each step, builds a table to show correspondences between each command and related H\W component.

Data Integration

This functionality integrates homogeneous data retrieved using the Knowing functionality, to create new data that is not explicitly present in the associated data source. In this specific case the integration rules depend on particular data correlation required by the user. DATA INTEGRATION involves the following actions:

- selecting typology of information starting from a topic and a particular keyword (for example Keyword = results of an Topic = experiment)
- accessing the Metadata Server (using Knowing)
- accessing the available sources (using Knowing)
- browsing data using Integration Criteria
- displaying integrated Information related to the topic.

Scenario: a scientist wants to compare results of experiments Exp1 and Exp2, on a given phenomenon. As example, ULISSE retrieves information on the ISS's position (Exp1) and Exp2 data or accelerometer data (Exp1) and Exp2 data. Then ULISSE integrates the data to allow the scientist to compare them using a browsing interaction.

Browsing

This functionality creates different schemas for browsing information retrieved using the Knowing functionality, in order to help the user to identify the data subset of interest. A set of data selection rules is available. BROWSING involves the following actions:

- accessing the Metadata Server
- selecting specific parameters
- composing a schema to browse data
- accessing the available sources
- exporting data
- browsing data according to the selected schema

Scenario: a scientist wants to retrieve a data subset of the experiment EXP on a given phenomenon, selecting a specific period of time, using a particular schema. ULISSE retrieves a subset of information regarding EXP on the basis of the parameters selected by the scientist and displays the data according to the chosen schema.

4.3.3 Example of a Searching and Browsing Session

The example session refers to the case of a user who wants to search and browse information related to the PULSAR2 experiment. Figures 4.8–4.11 show the screenshots from the ULISSE graphical frontend that has been developed by the Space Biology group at ETH Zürich. The scenario can be summarized in the following steps:

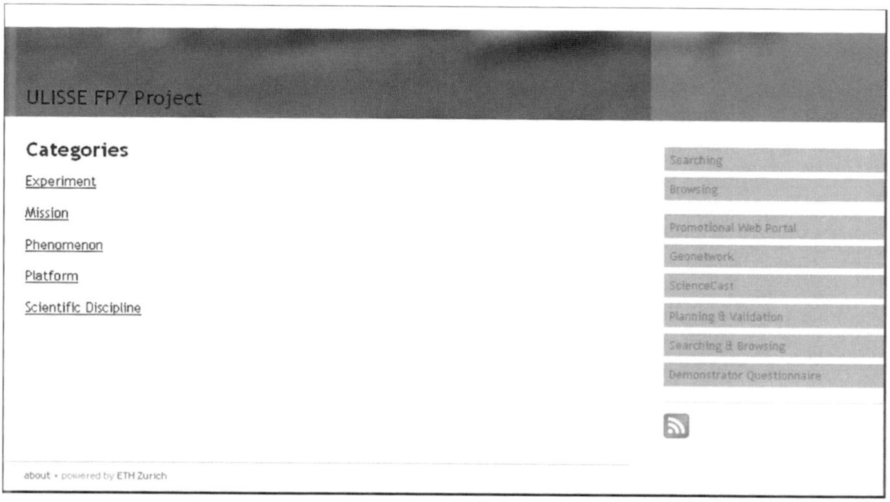

Fig. 4.8 List of categories

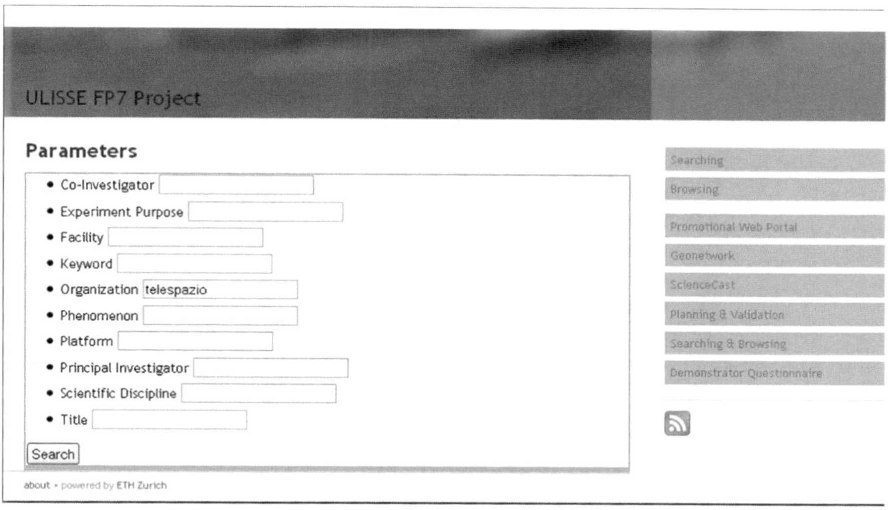

Fig. 4.9 List of parameters

- Selection of the category of interest ("Experiment" in this case), Fig. 4.8.
- Entering of values to retrieve the elements from the ULISSE domain. Figure 4.9 shows the ULISSE search page for the category "Experiment". According to this scenario, the user enters "Telespazio" in the text box corresponding to the "Organization" field;
- Display of the list of retrieved experiments.

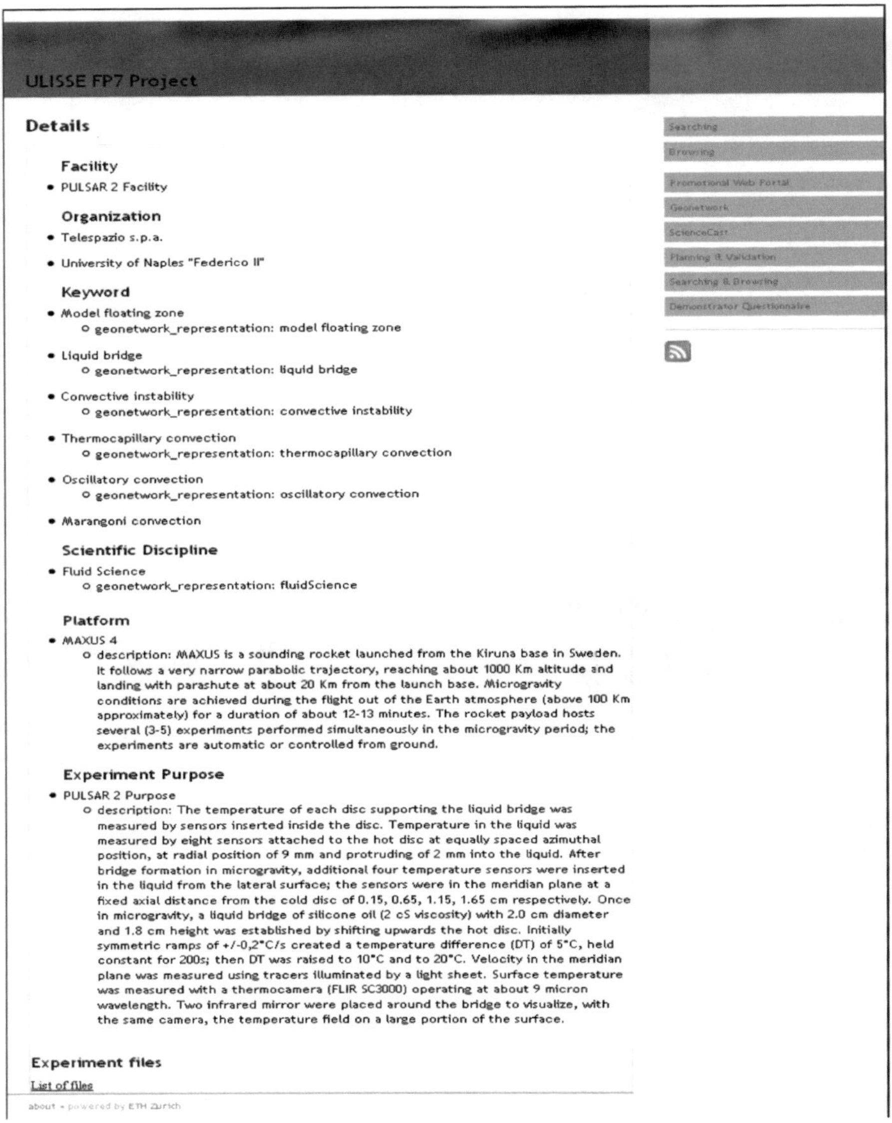

Fig. 4.10 Result of experiment browsing

- Selection of PULSAR2 experiment. The system browses the information (metadata) stored in the Knowledge Base related to the PULSAR2 experiment (Fig. 4.10). The list of available data can be accessed via the "List of Files" link. The user can choose a file from the list but only after authentication (Fig. 4.11), can he download the file.

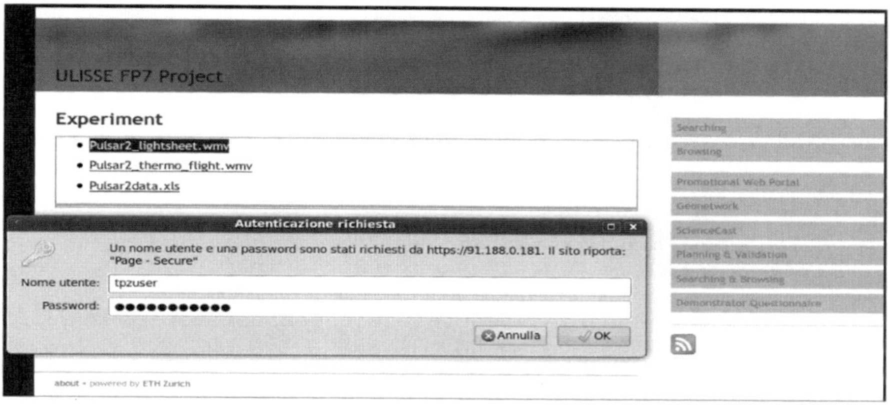

Fig. 4.11 Authentication request

References

1. CCSDS – Consultative Committee for Space Data Systems. http://www.ccsds.org/
2. Damen, D., Patriksson, M.: Putting Topic Maps to REST. In: Maicher, L., Garshol, LM. (eds.) Information Wants to Be a Topic Map. Proceedings of the Sixth International Conference on Topic Maps Research and Applications (TMRA'10), University of Leipzig and Leipziger Informatik-Verbund (LIV). ISBN 978-3-941608-11-5 (2010)
3. Damen, D., Pinchuk, R., Fontaine, B.: Creating topic maps ontologies for space experiments. In: Maicher, L., Garshol, LM. (eds.) Linked Topic Maps. Fifth International Conference on Topic Maps Research and Applications, TMRA 2009 Leipzig, 12–13 Nov 2009 Revised Selected Papers. Leipziger Beiträge zur Informatik. ISBN 978-3-941608-06-1 (2009)
4. ERASMUS Experiment Archive (ESA). http://eea.spaceflight.esa.int/
5. ESA – European Space Agency. http://www.esa.int
6. ESA – Human Spaceflight, Research and Applications Programme Board/Human Spaceflight Data Policy – ESA/PB-HSR(2004)29, rev. 2, corr.1 (2004)
7. "ESA's latest IPR policy" – in ESA Bulletin - February 2003 at [5]
8. European User Guide to Low-Gravity Platforms (UIC-ESA-UM-0001). http://www.esa.int/SPECIALS/HSF_Research 1094 (2005)
9. Fielding, R., Taylor, R.: Architectural Styles and the Design of Network-Based Software Architectures. University of California, Irvine (2000)
10. Garshol, M.: Towards a methodology for developing topic maps ontologies. In: Maicher, L., Sigel, A., Garshol, LM. (eds.) Proceedings of the 2nd International Conference on Topic Maps Research and Applications (TMRA'06), pp. 20–31. Springer-Verlag, Berlin/Heidelberg (2006)
11. Garshol, M., Moore, G.(eds.): Topic Maps – Data Model. ISO/IEC JTC1/SC34 Information Technology – Document Description and Processing Languages (2008)
12. Grimbach, A., Wever, P., Schneider, S., Willnecker, R.: Knowledge preservation: a semantic approach to visualizing and reusing microgravity material science data. Presented at the international astronautical congress 2010, Prague, IAC-10-D5.2.6 (2010)
13. ISRDB – International Space environment utilization Research DataBase (JAXA). http://idb.exst.jaxa.jp

14. Kuijpers, EA., Carotenuto, L., Cornier, C., Damen, D., Grimbach, A.: The Ulisse environment for collaboration on ISS experiment data and knowledge representation. Presented at the international astronautical congress 2010, Prague, IAC-10-D5.2.2 (2010)
15. MICREX – Microgravity Data Archive in Glenn Research Center (NASA). http://microgravity.grc.nasa.gov/fcarchive
16. Mikhailian, A., Dalmas, T., Pinchuk, R.: Learning foci for question answering over topic maps. In: Proceedings of the ACL-IJCNLP 2009 Conference Short Papers, pp. 325–328. Suntec, Singapore (2009)
17. OAIS Reference Model – CCSDS 650.B-1 at [1] (2002)
18. OAO Rocket and Space Corporation Energia. http://www.energia.ru/
19. PAIMAS Reference Model CCSDS 651.0-B-1 and CCSDS 651.0-M-1at [1] 2004
20. Pinchuk, R., Dalmas, T., Mikhailian, A.: Automated focus extraction for question answering over topic maps. In: Maicher, L., Garshol, LM. (eds.) Linked Topic Maps. Fifth International Conference on Topic Maps Research and Applications, TMRA 2009, Leipzig, 12–13 Nov 2009 Revised Selected Papers. Leipziger Beiträge zur Informatik. ISBN 978- 3-941608-06-1 (2009)
21. RKA's Scientific and Technical Advisory Council (STAC). http://knts.tsniimash.ru
22. Siegel, D.: Pull: The Power of the Semantic Web to Transform Your Business. Penguin Group. ISBN 978-1-59184-277-4 (2009)
23. http://www.ulisse-space.eu

Index

A

ACES. *See* Atomic Clock Ensemble in Space (ACES)
Advanced resistive exercise device (aRED), 150
Age-related extent of adaptive process, 126
Airborne particulate, 154
Air revitalization system (ARES), 154
Allergy, 152
Alloy, 16, 73, 83, 85–96, 183
ALTEA, 23, 24
Alteino, 23
Analysing interferometer for ambient air (ANITA), 154
Animal physiology, 4, 45–46, 123–129, 183
ANITA. *See* Analysing interferometer for ambient air (ANITA)
Antarctica, 142
Anti-g suit, 150
Anxiety, 141–143
API, 194, 195
aRED. *See* Advanced resistive exercise device (aRED)
ARES. *See* Air revitalization system (ARES)
Arrhythmia, 147
ASIM. *See* Atmospheric Space Interactions Monitoring Instrument (ASIM)
Association, 3, 4, 180, 187–190, 192, 194, 196
Astrobiology, 4, 12, 26, 28, 31, 103–107, 183
Astrobiology platform, 103–106
Atmospheric Space Interactions Monitoring Instrument (ASIM), 28
Atomic Clock Ensemble in Space (ACES), 27, 28, 60
Atom interferometry, 56
ATV. *See* Automated transfer vehicle (ATV)
Automated transfer vehicle (ATV), 30, 154
Availability, v, vii, 33, 43, 56, 82, 112, 119, 122, 140, 143, 153, 172, 175, 183, 184, 199

B

Back pain, 147, 151, 152
Bacteria, 28, 106, 112, 138, 156, 157
Basic functionality, 198, 200–202
Bed rest, 7, 34, 37–38, 45, 136, 140, 173
Bio-contamination, 154
Biofilm organisms surfing space (BOSS), 106
Biofluid, 75
Biolab, 11–13, 17
Biological life support systems (BLSS), 153, 156, 157
Biological rhythm, 139
BIOMEX, 106
Bion, 30–31
BIOPAN/FOTON, 31, 103–106
Bio-processing, 44–45
Bioregenerative life support, 156–157
Biotechnology, 11, 20, 30, 35, 47, 183
Bisphosphonate, 135
Boiling, 65, 68–69
Boiling crisis, 68–69
Bone, 2, 18, 23, 25, 31, 37, 40, 44, 82, 107, 110, 127, 130, 133–136, 139, 140, 145–147, 151, 152
Bone cell, 44, 110, 135
Braslet, 149, 150
Browsing, 195, 202, 204
Bulk metallic glass, 86–87, 93, 94, 96
Bull frog, 125, 126
Burn, 31, 68, 148

DOI 10.1007/978-3-642-21144-7,

C

Calcium, 120, 139, 147
Capillary flow, 70
Cardiac function, 139
Cardiomed, 24–25
Cardiovascular system, 25, 145
Casting, 82–84, 86, 88, 94, 95
CCSDS, 174, 181
Cell biology, 107–112, 123
Central nervous system (CNS), 23, 24, 115, 116, 130, 132, 133
Centrifuge, 12, 16, 17, 39–41, 46, 110, 124, 128, 149, 151
CEVIS. *See* Cycle ergometer with vibration isolation and stabilization (CEVIS)
Chibis, 149, 150
Clinostat, 42–43, 46, 108, 123, 128
CNS. *See* Central nervous system
CO_2, 8, 74, 146, 152–154, 156, 157
Cognitive function, 130, 142, 143
COL-CC, 176, 177
Columbus, v, vi, 2, 10–20, 24, 26–30, 90, 98, 131, 132, 154, 176, 177, 181
Columnar, 87–89, 94, 95
Columnar-equiaxed, 87–89, 94
Combustion, 4, 11, 20, 22, 35, 56, 69, 75–81, 148, 172
COMEX's Hyperbaric Experimental Centre, 39
Computer-aided diagnosis and surgery, 151
Confidentiality, 199, 200
Confinement, 38–39, 138, 141, 142, 144, 146, 151, 153
Constraint language, 187, 191–192
CO_2 sensitivity, 152
Countermeasure, 18, 25, 31, 37, 38, 40, 115, 117, 123, 129–131, 133, 135, 140, 146, 148–152
Crew surgeon, 146
Critical point, 22, 56, 65–68, 72, 74
Crystal, 14, 20, 24, 31, 49, 56, 59, 61, 70, 72, 84–87, 89, 93–95, 112, 183
Cu-based alloy, 94
Custodian, 3, 175, 177, 178, 183
Cycle, v, 4, 27, 43, 97, 98, 113, 121, 122, 149, 154, 171, 174, 176–177, 182
Cycle ergometer ?VB–3?, 149
Cycle ergometer with vibration isolation and stabilization (CEVIS), 149, 150

D

Data access, 4, 181, 194–205
Database, 143, 172–173, 177, 178, 180, 182, 192, 196, 198
Data description, 171–185
Data format, 65, 181–182, 191, 193
Data integration, 4, 201–202
Data model, 186–192
DEBIE–2, 27
DECLIC, 21, 22
Depression, 125, 139
Diffusion, 13, 21, 31, 49, 65–67, 70–73, 75, 76, 78, 83, 112
Directional solidification (DIRSOL), 21, 22, 84
DIRSOL. *See* Directional solidification (DIRSOL)
Discipline, 1, 3, 4, 10, 11, 14, 96, 97, 103, 186, 192, 196
Documentation, 174, 175, 178, 180–183, 185, 195
Dosimetric Telescope (DOSTEL), 24, 27, 102
DOSTEL. *See* Dosimetric Telescope (DOSTEL)
Drop towers, 35–36, 65, 77, 124
3D-SPACE, 132
Dust, 50, 56, 61–63, 79, 98
Dust cloud, 79

E

Earth viewing camera (EVC), 27
EDR. *See* European Drawer Rack (EDR)
Einstein's equivalence principle, 58
Elaborator of televised images in space (ELITE-S2), 25
Electrolyte level, 139
Electrolytic process, 156
Electromagnetic, 8, 16, 27, 83, 89, 91, 92, 97, 99–101
Electro-magnetic levitator, 14, 16, 30, 89–92
EMCS. *See* European Modular Cultivation System (EMCS)
Emulsion, 13, 16, 49, 59, 69, 71–72
Endothelial cell, 109–110, 128
ENFORM. *See* European Network using Fish as Osteoporosis Research Model (ENFORM)
Environmental data, 179
Epithelial cell, 108
EPM. *See* European Physiology Module (EPM)

Ergometer, 149
ESA data policy, 184, 199
ETD. *See* Eye tracking device (ETD)
Ethical issue, 183–185
European Control Centre Network, 28–30
European Drawer Rack (EDR), 11, 14, 16, 17, 30
European Modular Cultivation System (EMCS), 11, 17, 46
European Network using Fish as Osteoporosis Research Model (ENFORM), 127, 128
European Physiology Module (EPM), 11, 14–15, 18, 24, 132
European Physiology Modules Facility, 11, 18, 24
European Technology Exposure Facility (EuTEF), 12, 26, 102, 103
EuTEMP, 27
EVC. *See* Earth viewing camera (EVC)
Experiment sponsor, 175
EXPOSE, 27, 28, 103, 104, 106
EXPOSE-E, 103, 104
EXPOSE-R, 28, 103, 104, 106
Extreme environment, 39, 47–48, 104, 129, 145
Eye, 18–19, 98, 126, 131–133, 147
Eye tracking device (ETD), 18–19

F
Facility for Adsorption and Surface Tension Studies (FASTER), 14, 16
Fatigue, 116, 133, 140, 141, 144
Fe-based alloy, 86, 94
FIPEX, 27
Fish, 123, 125–128, 183, 192
Flame spread, 79–80
Fluid, 4, 10, 11, 13–14, 20, 22, 24, 31, 35, 37, 41, 47, 49, 56, 59, 61, 64–75, 81–84, 89, 91, 95, 108, 112, 139, 140, 146, 147, 149, 150, 152, 172, 179, 192
Fluid balance, 139
Fluid Science Facility, 49
Fluid Science Laboratory (FSL), 10, 13–14, 179, 181
Fluid shift, 37, 139, 146, 147, 149, 150, 152
Flywheel exercise device (FWED), 18, 19, 37
Foam, 10, 49, 59, 65, 66, 69, 71, 96
Food, 16, 139, 155, 156
Format, 65, 172, 178, 181–182, 186, 191, 193
Foton, 30–31, 72
Free fall machine, 42–43
Frontend, 202
FSL. *See* Fluid Science Laboratory (FSL)
Fuel droplet, 78–79
Fuel particle, 79
Functionality, 28, 44, 132, 198, 200–202
FWED. *See* Flywheel exercise device (FWED)

G
Gas contaminant, 154
Gaseous diffusion flame, 78
Gas sensor, 154
Gastrointestinal disorder, 147
General relativity theory, 57, 59
Giant fluctuation, 72
Glovebox, 11, 12, 17–18, 20–21, 182
Granular matter, 59, 63–64
Gravitational cell biology, 107–112
Gravitational mass, 57
Ground-based Facility, 37–50, 65, 77, 172
Groupthink, 145

H
Hand grip dynamometer, 25
Headache, 146
Health prevention, 148–151
Hearing, 152
Heart rhythm, 139
Heat transfer, 10, 65, 68–69, 80
Heavy ion radiobiology, 115–117
High-altitude balloon, 33, 36–37
High efficiency particulate air (HEPA) filters, 154
High energy, 101–103, 111, 113
Housekeeping, 2, 21, 65, 177, 179
HRDP, 177, 178
HR-pQCT, 136
Human centrifuge, 40–41
Human performance, 142, 143
Human physiology, v, 18, 40, 123, 129–141, 175, 182–185, 192
Human-rated linear and angular accelerator, 41–42
Hyperbaric chamber, 38, 39, 142
Hypergravity, 39, 40, 110, 119, 120, 124, 125, 129
Hypokinesia, 37–38

I
Immune system, 116, 127, 128, 136, 138, 140, 141, 146, 152
Immunological system, 136–137
Inertial mass, 57

Infectious disease, 152
Integrative physiology, 139–140
Integrity, 3, 135, 139, 199
Interface, 4, 14, 16, 22, 44, 49, 69–72, 74, 95, 97, 174, 177, 181, 182, 194, 196
Interfacial transport, 70–71
International Space Station (ISS), v–vii, 2–4 7–30, 33, 59, 61, 62, 65, 77, 81, 90, 98, 100–107, 121, 131, 132, 134–136, 138, 141, 147, 149–151, 154, 155, 173, 175, 176, 178, 196, 200
Interpersonal relationship, 143–145
ISO 19115, 181, 186
ISO 19139, 181
ISS. *See* International Space Station (ISS)

K

Karkasse, 149
Kidney stone, 147
Knowledge, vii, viii, 1–4, 23, 38, 59, 61, 74, 76, 79, 80, 89, 94, 106, 107, 125, 129, 132, 139, 141, 173, 181, 185–196, 204
Knowledge base, 3, 4, 181, 185–196, 204
Knowledge domain, 185, 187, 190, 192
KUBIK, 14, 16, 17

L

Legal aspect, 4, 183–185, 200
Level, v, 7–9, 12, 23, 30, 34, 43, 45, 48, 56–59, 61, 66, 80, 82, 86, 89, 93, 101, 106, 107, 115, 119–122, 125–128, 130, 132, 134, 139–141, 144, 146, 148, 150, 153–155, 173, 176–183, 186, 199
Life support system (LSS), 4, 38, 44, 47–140, 153–157
Liquid bridge, 66, 70
Long Duration Exposure Facility (LDEF), 103
Low body negative pressure (LBNP), 149, 151

M

Magnesium, 75, 140
Magnetic levitation, 46–47, 65, 89, 91, 108, 124
Magnetic resonance, 45, 46, 132
Manipulation, 140, 142
Marangoni, 65, 66, 70, 72, 201
MARES. *See* Muscle Atrophy Research and Exercise System (MARES)
Mars 500, 38
Matching, 65, 201
Material, v, 4, 8, 10–12, 16, 20–23, 26, 27, 31, 33, 35, 39, 41, 46–50, 59, 62, 70, 71, 73–76, 80–98, 103–105, 107, 115, 122, 155, 157, 183, 192, 194
Material Science Laboratory, 11, 21–22, 91
Material Science Research Rack (MSRR–1), 11, 21–22
Materials Exposure and Degradation Experiment orbit (MEDET), 27
Materials Science Facility, 49, 50
Matroshka, 22, 23, 102
MCC. *See* Mission Control Centre (MCC)
Mechano-perception of cell, 109
Medaka, 123, 127
Medes Clinical Research Facility, 38
MEDET. *See* Materials Exposure and Degradation Experiment orbit (MEDET)
MELFI, 25–26, 138
MELiSSA. *See* Microbial Ecological Life Support System Alternative (MELiSSA)
Mental disorder, 139
Metadata, 3, 4, 172, 181, 183–186, 192–196, 199–202, 204
Metadata standard, 3, 181, 185–186, 192–195, 199
Metallic glass, 84, 86–87, 93, 94, 96
Mice drawer system, 25
Microbe, 104, 105, 112, 138
Microbial Ecological Life Support System Alternative (MELiSSA), 47, 156, 157
Microfluidic, 75
Microgravity data, 179
Microgravity science glovebox, 11, 20–21, 182
Miller-Urey Experiment in space (MUE), 106
Mission Control Centre (MCC), 2, 25, 29, 175–177, 182
Monkey, 126, 183
Motor skill, 142–143
Movement analysis, 45
Multi electrodes encephalogram measurement module (MEEMM), 14, 132
Multispectral plant imaging, 46
Muscle, 2, 11, 14, 18, 37, 45, 107, 126, 127, 129, 131, 132, 135, 140, 145, 146, 151, 152
Muscle Atrophy Research and Exercise System (MARES), 11, 18
Musculoskeletal trauma, 148
Myoelectrostimulation, 150

N
NASA Human Research Facility, 11
Neuroendocrine system, 136
Neuroscience, 131–133
Ni-based superalloy, 93–94
Nutrition, 37, 38, 45, 123, 129, 135, 138–141, 149

O
OAIS, 174
Occurrence, 143, 189–190, 194, 195
Ontology, 190, 192, 194, 199–201
O/OREOS, 105
Operational space medicine, 145–152
Optical clock, 56
OREOCUBE, 107
Orthostatic intolerance, 124, 133, 149
Orthostatic tolerance, 140
Owner, 102, 183, 199
Oxygen production, 155
Oxygen supply, 154

P
PAIMAS, 174
Parabolic flight, 33–35, 65, 89–92, 111, 120, 124, 149, 182
PEMS. *See* Percutaneous electrical muscle stimulator (PEMS)
Penguin suit, 149
Perceptual skills, 142–143
Percutaneous electrical muscle stimulator (PEMS), 18
Phase, 2, 16, 21, 24, 32, 34, 56, 61, 65–74, 78–80, 82, 83, 85, 86, 90, 91, 93–95, 106, 128, 144, 148, 153, 156, 157, 175–176, 179–182
Phase transition, 56, 61, 65–67, 69, 70, 74, 83
Photochemistry on the Space Station (PSS), 106
Physico-chemical air revitalization, 154
Physico-chemical life support system (PhLSS), 153
Plant, 4, 11, 12, 17, 28, 40, 42, 46, 47, 93, 117–122, 155–157, 173, 183, 192
Plant Biocentre, 46
Plant gravitropism, 42, 117–119, 122
Plant gravity sensing, 117–119
Plasma, 8, 24, 27, 41, 50, 56, 59–62, 64, 99–103, 118, 139
Plasma Electron Gun Payload (PLEGPAY), 27
Plasma kristall, 24
Player, 187–189
PLEGPAY. *See* Plasma Electron Gun Payload (PLEGPAY)
Policy, 4, 103, 112, 175, 178, 184, 196, 199, 200
Portable pulmonary function system, 15, 16
Posture, 25, 42, 130, 131
Premixed gas flame, 77–78
Preservation, 2, 171–175, 182, 183, 186, 196
Pressure, 8, 13, 14, 22, 24, 36–39, 50, 63, 65–67, 69, 71, 72, 77, 79, 94, 118, 128, 129, 133, 144, 145, 149, 152–154, 179
Principal investigator, 81, 173, 175, 178, 193
Processing level, 178–181
PSS. *See* Photochemistry on the Space Station (PSS)
Psychology, 38, 45, 141–145
Publication, 2, 10, 117, 124, 172, 176, 179–182, 185, 194, 195
Pulmonary disorder, 147
Pulmonary function system, 15, 16

Q
Quantum communication, 56
Quantum gas, 56
Quantum physic, 56–57
Quantum theory, 57
Query language, 187, 191
Quintavre, 149

R
Radiation, 1, 4, 7, 8, 23–24, 26–28, 31, 33, 48–49, 97–104, 106, 107, 113–117, 122, 128, 130, 138, 141, 146, 151–153
Radiation biology, 4, 104, 106, 113–117, 128, 138
Radiation monitoring facility, 23–24
Random positioning device, 42–43
Random positioning machine, 43, 108, 121, 123
Raw data, 176, 179
Rehabilitation, 135, 151, 152
Relativity, 28, 57–59
Renal stone, 140, 147
Representational state transfer (REST), 195
Resource, 3, 10–12, 14, 21, 43, 48, 76, 81, 82, 128, 153–155, 181, 182, 195–199
Respiratory system, 147
REST. *See* Representational state transfer (REST)
Reverse osmosis, 156

Role, vi, 3, 12, 23, 28, 39–42, 64, 65, 69, 75, 108, 109, 111, 118, 121, 133, 139, 140, 185–189, 192, 197–199
Rotating wall vessel (RWV), 123, 124, 126–128
Russian system for water recovery from humidity condensate (SRV-K), 155

S
Sample, 12, 16, 21, 22, 26, 28, 41, 47, 49, 65, 67, 73, 74, 90–92, 95, 104, 106, 107, 111, 121, 138, 154, 178, 179, 182–183, 194, 195
Scenario, 3, 80, 115, 153, 155, 172, 198, 200–203
ScienceCast, 187
Scientific data, vi, 2, 27, 172–174, 176–179, 182–184
Scientific investigator, 2, 175, 183
Scope, 59, 61, 63, 185, 190, 192, 200
Security, vii, 2, 76, 178, 181, 196, 199
Semantic technology, 185–186
Sensorimotor system, 130–133
Service, 3, 4, 26, 29, 31, 33, 34, 36, 43–45, 83, 94, 98–102, 173, 183–185, 187, 189, 194–205
Session, 149, 150, 176, 181, 202–205
Short arm centrifuge, 151
SilEye–3, 23
Simulation Facility for Occupational Medicine Research (AMSAN), 38
Skeletal system, 133–136
Skin, 23, 98, 131, 147, 152, 192
Sleep, 14, 133, 141, 143, 144, 146–148
Sloshing, 73
Smouldering combustion, 80
Snail, 125, 126
Social interaction, 141
SODI, 21, 182
Soft matter, 56, 59, 61, 83
SOLAR, 12, 27, 98
Solution Crystallisation Diagnostics Facility (SCDF), 17
Sounding rocket, 32–33, 36, 65, 74, 77, 89–90, 95, 98, 108, 111, 120, 176
Source, 3, 33, 48, 49, 75, 85, 94, 113, 115, 116, 141, 148, 153, 156, 157, 181, 183, 198–202
Space adaptation syndrome, 149
Space agency, v, 2, 9, 10, 28, 30, 35, 37, 90, 112, 175, 179, 201
Space anemia, 140
Space environment, vii, 1, 4, 7–9, 26, 28, 38, 44, 48, 65, 97, 100, 103, 104, 106, 107, 111, 123, 140–143, 153, 172
Space infrastructure, 2, 9–31
Space motion sickness, 130, 131, 146, 152
Spine, 134, 152
Spray, 78–79, 96
Statocyte, 118
STONE experiments, 107
Superalloy, 87, 93
Supercritical fluid, 66–68, 74

T
Tail or hind limb suspension, 124
Tail suspension, 124, 126
Telemedicine, 43–44
Telemetry, 30, 36, 37, 100, 177–179, 181, 182
Temperature, 7, 8, 16, 22, 24, 26, 31, 33, 37, 48–50, 63, 65–68, 70, 72–74, 76–82, 84–87, 89–94, 96, 98, 105, 106, 128, 139, 154, 179
Thermalization, 67
Thermodiffusion, 73, 74
Thermophysical, 83, 85, 89, 93, 94
Thermo-soluto-capillary, 70
Thigh cuff, 149
THLD. *See* Transient heel loading device (THLD)
Ti-alloy, 87, 93
Tissue engineering, 44–45, 124
Toad, 125
Toadfish, 125, 126
Topic, 56, 57, 61, 63, 128, 171, 179, 181, 183, 186–196, 199, 201, 202
Topic map, 186–195
Topic type, 187, 194
Transient, 73–74, 79, 142, 151, 152
Transient heel loading device (THLD), 151
Trauma, 148
Treadmill, 149–151
Treadmill with vibration isolation and stabilization (TVIS), 150
TReK, 182
TRIBOLAB, 27
Two-phase flow, 68–69

U
Unicellular protist, 110–111
Universality, 58, 59, 61, 66, 67

Urine processor assembly (UPA), 155
Urine retention, 147
User, vii, viii, 2–4, 12–15, 17, 18, 20, 22, 29, 30, 33, 44, 46, 90, 100, 173, 174, 176, 178, 181, 184–187, 194, 195, 197–204
User Support and Operations Centre (USOC), 2–4, 14, 15, 17, 20, 29–30, 46, 172, 174–180, 182–184, 195, 196, 199
USOC. *See* User Support and Operations Centre (USOC)
UV solar radiation, 97–99

V
Vacuum, 1, 8, 12, 16, 21, 26, 28, 36, 48, 107, 149
Vessel identification system, 19–20
Vestibular sensitization, 125, 126
Vibration, 8, 21, 35, 63–64, 66, 70, 73, 74, 83, 141, 149–151
Vision, vi, 131, 147, 152
Vitamin D, 98, 127, 139
Vitamin E, 140
Vitamin K, 140
VITRINE, 106

W
Water, 16, 17, 22, 31, 47, 66, 68, 69, 71, 72, 139, 147, 150, 154–157
Water electrolyser, 154
Water processor assembly (WPA), 155
Water recovery, 155
Well-being, 141–146

X
XtremeCT, 136

Y
Yeast, 13, 111–112

Z
Zebrafish, 123, 124, 126–128

The views expressed in this book are those of the ULISSE project and do not necessarily represent those of the European Commission.

ULISSE is a Collaborative Project funded under the EUs Seventh Research Framework programme.

D. Beysens et al. (eds.), *Laboratory Science with Space Data*,
DOI 10.1007/978-3-642-21144-7, © Springer-Verlag Berlin Heidelberg 2011